THE MODERN MYTH
Ancient Astronauts and UFOs

OTHER BOOKS by Morris Goran

Introduction to the Physical Sciences
Experimental Chemistry for Boys
Experimental Biology for Boys
Experimental Astronautics
Experimental Earth Sciences
The Core of Physical Science
Experimental Chemistry
The Story of Fritz Haber
Biologia Experimental
The Future of Science
Science and Anti-Science
A Preface to Astronomy

THE MODERN MYTH
Ancient Astronauts and UFOs

Morris Goran

South Brunswick and New York: A. S. Barnes and Company
London: Thomas Yoseloff Ltd

© 1978 by A. S. Barnes and Co., Inc.

A. S. Barnes and Co., Inc.
Cranbury, New Jersey 08512

Thomas Yoseloff Ltd
Magdalen House
136–148 Tooley Street
London SE1 2TT, England

Library of Congress Cataloging in Publication Data

Goran, Morris Herbert, 1916–
The modern myth, ancient astronauts and UFOs.

Bibliography: p.
Includes index.
1. Flying saucers. 2. Civilization, Ancient. 3. Man,
Prehistoric. 4. Interplanetary voyges.
I. Title.
TL789.G67 1978   001.9'42   76-50190
ISBN 0-498-02008-8

PRINTED IN THE UNITED STATES OF AMERICA

for
Marjorie and Harvey
as well as Ruth and Toby
and of course Cymia

Contents

THE MODERN MYTH
Ancient Astronauts and UFOs

I

Introduction to
Ancient Astronauts

During the first half of the twentieth century, American astronomers apparently followed the lead of Henry Norris Russell and publicly, at least, proclaimed the existence of one and only one solar system. There was no observational evidence to believe otherwise. Gradually, the principle of the uniformity of nature took hold, and some astronomers saw our sun and its attendants as typical rather than unique. Peter van de Kamp's multidecade study at Swarthmore College of Barnard's star showed a wiggle probably due to an accompanying planet or planets, and the more conservative had a bit of the required evidence for the existence of other solar systems. Inferences followed: some solar systems had planets capable of supporting life; and some life was more intelligent than man on earth.

Solar-system study became more important, too, during the 1950s because Americans and Soviets vied for the distinction of being first to have placed an artificial satellite around the earth. The success of the Soviet Union in 1957 was in a large measure responsible for the billion-dollar research and operation agency established in 1958 by the United States—the National Aeronautics and Space Administration. A race to be the first human being on the moon captured the attention of many Americans and received official sanction from a presidential statement in 1960.

In 1960, a young astronomer, Frank Drake, used the facilities of the National Radio Astronomy Observatory, Green Bank, West Virginia, to try to detect radio signals manufactured by an intelligence beyond our solar system. He studied two stars for a few hours during a period of two weeks, and the results were negative. This Project Ozma, embarrassing to some professional astronomers, gave new horizons to scientists, artists, and writers.

Life on other planets, and in other solar systems, a favorite topic of science fiction, became more respectable. Several Europeans easily crossed the barrier from imagination to reality and proposed so-called advanced forms of intelligence living elsewhere. The hypothesis seemed to receive approval from the suggestion, now forsaken, of Russian astrophysicist I.S. Shklovskii that the two small moons of Mars were artificial. In France, Jean Sendy as well as Jacques Bergier wrote about early visitations of earth, perhaps thousands of years ago, by extraterrestrial astronauts. Among those to embrace the theme was Erich von Däniken, a Swiss. His books proved to be most popular, bringing many adherents to the idea.

The Ancient Astronaut Society was organized in 1973 by a Park Ridge, Illinois, attorney. They had their first World Conference on Ancient Astronauts for about three hundred attendants, April 1974, in a Chicago suburb. (Chicago newspapers had advertisements for the gathering under the heading of *amusements*).

The Second World Conference on Ancient Astronauts was in Zurich, Switzerland, May 1975. Several hundred members of the Ancient Astronaut Society met at the O'Hare Inn, Des Plaines, Illinois, March 1976. They had another midwinter meeting there February 26, 1977, but their Third World Conference was held in Yugoslavia, and the Fourth was in Rio de Janeiro in June 1977.

The number attending the conventions is not indicative of the mass appeal on the subject of ancient astronauts;

Von Däniken's books, for example, have sold through June, 1977 a total of thirty-eight million copies.

American television hastened the popularity of the ancient astronaut theme. In 1972, the National Broadcasting Company had a one-hour special program called "In Search of Ancient Astronauts." The show was so successful that in December 1973 they put together another—"In Search of Ancient Mysteries." The producer had a paperback book with the same title released to coincide with the showing.

The following chapters analyze the ancient astronaut theme and the alleged supporting evidence. Erich von Däniken is only one of the authors examined.

1

The Roots of Civilization

Many of those who promulgate the ancient astronaut theme seem to present earliest men and women on Earth as incompetent brutes. In his first book distributed in the United States, von Däniken charges: "Let us not forget that we too were semisavages 8,000 years ago."[1] In his second book he reports: "For the 'savages' who painted these memories of the visit of the 'gods' on the rock faces were probably at the mental age of a child."[2] The lesser-known in the camp have echoed the calumny. Eric Norman argues:

> Let's look at what history tells us of the origin of civilization. One day about seven or eight thousand years ago, according to the historians, our ancestors were living in caves and clubbing their way through a dangerous brutal world. Without prior notice, a miracle occurred in the Mesopotamian Valley that was the origin of our civilization. Virtually overnight, a group of primitive tribesmen were transformed into skilled, educated citizens of Sumerian city-states. We are told that the Sumerian cities sprang up spontaneously.[3]

Alan and Sally Landsburg give a more sophisticated view of the contention by citing that the artistic, mathematical,

14

and musical abilities of human beings had no survival value, and some other way had to be found to explain the development of those talents.[4]

A couple of the authors promoting extraterrestrial visitations ages ago have a more acceptable view of early peoples on earth. Andrew Tomas sets forth an aim "to demonstrate that the technical skills of the men of antiquity and prehistory have been greatly underestimated."[5] Jean Sendy is most explicit:

> Thus by 22,000 B.C. art was already organized. The people of that time lived in huts and tents. They wore clothes made of finely sewn skins. They adorned themselves with necklaces and other ornaments made of animal teeth, shells and carved bones. They knew how to make baskets and work with wood and bark. They had flint tools, shaped according to their uses, and bone instruments. They were skilled butchers and furriers. And they had religious convictions, as is shown by the fact that skeletons from that period have been found arranged in a way that could not have been fortuitous, or surrounded by objects indicating conclusively that death was an occasion for ritual.[6]

Sendy also gives the earliest people credit for knowing the phenomenon of precession of the equinoxes—a feat usually ascribed to the Father of Greek astronomy, Hipparchus.[7]

The facts are that not fully bipedal, small-brained near-men or men-apes knew how to make and use tools. Neanderthal man, living anywhere from 600,000 to 30,000 years ago, had weapons against cave animals, used paints, had flint materials, and buried their dead. One expert claims that "it is obvious that they did not lack inventive powers."[8] Friedrich Klemm credits the Paleolithic or Old Stone Age with "the axe, the flint scraper, the spear, bow and arrow, gimlet, oil-lamp and all sorts of bone instruments," as well as the first trap for big game, essentially a lever mechanism.[9] During the Neolithic or New Stone Age, men and women engaged in pottery, carpentry, weaving, baking, brewing, and smelting of copper; they had the potter's wheel, wagon wheel, hammer, hoe, tents, rafts, and boats. Early peoples are responsible for the momentous discovery and use of the wheel, fire, and metallurgy; in due time they learned how to domesticate plants and animals.

The ancient astronaut theorists, generally unaware and unappreciative of the accomplishments of early mankind, stand in awe of some discoveries about them. The writers would rather assign artifacts and buildings to the acumen of beings outside the planet, who visited and bestowed a cornucopia upon the earthly dimwits. Andrew Tomas describes this conception as given in the Babylonian tradition: "Dwellers in the Euphrates valley were beast-like before Oannes, but after him they became civilized and reached a high level of intellectual development."[10] Oannes, with the head of a fish covering a human head, came out of the Persian Gulf and is interpreted to have been a superior extraterrestrial being; the fish's head was really the early people's way of telling about the space helmet.

The writers generally agree that the ancient astronauts arrived on Earth by air. Von Däniken was one of the first to cite the array of lines near the city of Nazca in the Peruvian Andes as an airfield built by extraterrestrial visitors. Marie Reiche, a scientist who has studied the Nazcan figures for many years, explains that the pre-Inca people responsible for the geometric designs and drawings worked them out first on small plots still seen near many of the larger figures. At a May 1974 symposium at the University of California, Berkeley, a teaching assistant in anthropology showed how the lines, seen from nearby mountains, could have been drawn first and then projected onto the plain through elementary surveying methods. He reported the layout to be harmonious with Nazca pottery designs, as well as Peruvian thought.

Careless tourists have now destroyed large sections of

the Nazca lines, about 250 miles south of Lima, Peru. Through car tracks and footprints, thousands of sightseers have obliterated some of the furrows not far from the Pan-American Highway. The Peruvian government's National Institute of Culture has started a protection plan, with guards, guides, an observation tower, and a periodic fly-over program. However, parts of the drawings of the spider, the monkey, and the big bird are already destroyed.

The water conduits of precisely fitted stones at the city of Tiahuanaco, South America, are, for von Däniken, a marvel unattainable by Earthly people without extraterrestrial help. He and others are amazed at the Inca or pre-Inca fortress at Sacsahuaman, where several hundred large blocks are fitted neatly together; the smallest may be two tons, and the largest may be about one hundred tons. On pages 355–6 of his 1974 book, *In Search of Ancient Gods*, von Däniken describes the 400,000 basalt blocks on Nan Madol, a tiny island in the Carolines, and reports: "One of the islander's legends relates that a flying dragon helped with the massive transport job."

Early peoples were evidently expert stoneworkers. Moreover, unlike recent technological societies that accent planned obsolescence for practically all products, the communities of ancient times built structures as lasting monuments. Those in South America are indicative of others elsewhere, with the techniques of construction still a puzzle to modern engineers; for example, about 525 B.C., Eupalinus supervised the digging of a tunnel out of solid limestone in order to carry water to the city of Samos on the island of Samos off the coast of Ionia in the eastern Mediterranean. The tunnel, about 3,400 feet long, is straight; how this was accomplished is yet to be solved.[11]

Alternatives to "solving" the problem with extraterrestrial visitations include the exotic thesis that some early peoples, whose societies were destroyed by a natural catastrophe, lived at a level we have not yet reached.[12] The simplest way out, however, is to realize the genius of our predecessors. Professor Bernard Wailes, anthropology department, University of Pennsylvania, wrote: "The monuments reveal architectural and engineering skills of a high order, particularly since most were built without the benefit of a metal technology. Each new careful excavation, revealing further details of these ingenious structures, adds to our respect and admiration for the men who fashioned them."[13]

A recent study of American Indian architecture by Professor Ralph Knowles of the University of Southern California shows that the pueblos in the Southwest were carefully designed for efficient use of solar energy and winds in the area. The choice of materials and method of construction maximized sun entry in the winter and minimized it in the summer.

It is known that man has a vast heritage in making and using tools, an inheritance stretching into organisms less specialized. Singer, Holmyard, and Hall cite a variety of finch in the Galapagos Islands that uses a cactus spine or twig held in its beak to poke out insects imbedded in trunks of trees.[14] Wasps have been known to use pebbles to tamp the soil over nest entrances. Southern sea otters place shellfish on a slab of rock and open the organism by pounding it with a stone. A recent report claims that laboratory-raised Northern blue jays were observed tearing pieces of newspaper to use as tools for raking in food pellets otherwise out of reach, a feat generally attributed to monkeys.[15] For several years Jane Goodall observed toolmaking and tool-using among wild chimpanzees and frequently saw them make probes out of twigs by stripping off the leaves. The probes were pushed into holes in termite nests, and the termites on the probe were eaten. Could not man with a greater ability to solve problems have been able to fashion strong tools from hard rock? In at least two cases—the monuments on Easter Island and the Pyramids in Egypt— enough evidence is available to dispute effectively the contention that man had to be taught by a superior being from elsewhere.

Von Däniken presents the giant stone faces on Easter Island as work that had to be done by extraterrestrial intelligences. Why they would want to engage in this activity is not his concern; he only calculates the accomplishment as a superhuman effort. Explorer Thor Heyerdahl gives a directly contradictory opinion, formulated before von Däniken reached his. Heyerdahl reports that only a few men using stone axes would be needed to carve the giant faces, and the kind of stone axes to be used are on the island. Moreover, about one thousand men would have been enough to take the statues from the quarry, and about half the number would be needed to move the sculptures to other parts of the island. Heyerdahl showed that the stone statues could be moved across the countryside of Easter Island by having 180 men actually do so.[16]

Von Däniken's description of the Easter Island statue size, between thirty-three and sixty-six feet high, is at variance with what other authors claim. Paul Wingert, on page 314 of *Primitive Art,* published in 1962, describes the Easter Island work as being from "about 12 to over 30 feet high." Alberto Ambesi, in *Oceanic Art,* published by Hamlyn Publishing Group, reports on page 128 that the statues range from nine to thirty feet.

A young archaeologist, Edmundo Edwards, who spent fourteen years on Easter Island, lectured in Chicago early in 1975 and claimed that von Däniken, when talking to him, cast aside all accepted archaeological explanations and wanted to hear only things that could not be explained. Edwards showed him an octopus drawn on a rock where a native had later superimposed the head of a human being. Von Däniken presents this in his book *Gold From the Gods* as evidence of extraterrestrial visitation.

For religious, military, or economic reasons, huge stones were moved in many early societies. Anthropologist Robert

F. Heizer has documented many cases of great stones being moved, which weigh hundreds of tons. He, too, mentions the remarkable stone-joining at Sacsahuaman: "It aroused the interest of the Spaniards, and in response to their inquiries in the 16th century they were told that the stones were softened and thus made easy to work by application of the juice of certain red leaves."[17]

Facts to support the stone-softening contention are available in a book with a small amount of science and a great deal of pseudoscience.[18] An earthenware jug with a black, viscous fluid was found in a Peruvian burial ground. The liquid turned rocks into a soft putty. A kingfisherlike bird in the Bolivian Andes nests in holes bored out of rock; reports indicate that the bird prepares his house by rubbing a leaf on the stone until it is soft and manageable. Finally, a British explorer in Peru noted how a pair of large Mexican-type spurs were severely corroded after one day's contact with the juice from plants with red, fleshy leaves.

As far as is known, the Egyptians had no rock softeners, but their pyramids have been the most extensively studied of the early structures. Approximately seventy pyramids of significant size are in Egypt. The Cheops, or Great Pyramid, at Giza is the largest, with a bottom portion covering about thirteen acres. It was constructed from about 2,300,000 blocks, each weighing about 2½ tons. The number of men involved in building this colossus has been estimated to be anywhere from 100,000 to 250,000. The most popular theory to account for the construction is that a sloping roadway was built up to the level of operation. The rocker or cradle theory assumes that a stone was laid upon a solid wooden sled with a cylindrical undersurface, as if cut from a tree trunk, and this was rocked by great levers. The sand-mound theory postulates that as each layer of stones was placed it was surrounded by a bed of sand sloping downward to the surface of the ground; the materials for the next layer were dragged up this incline; the completed pyramid was buried in a mountain of sand, then dug away.

Much is yet to be uncovered about ancient Egypt and its society, so that the reason for pyramid construction is now hazy. Late twentieth-century scholarship has tended toward the belief that the construction crew were farmers awaiting the growth of crops, and the monuments were religious and political memorials. Perhaps the motivation was similar to

The Great Pyramid, with the Sphinx in the foreground

that for such edifices as the Colosseum, the Parthenon, the Palace at Versailles, and the Pentagon in Washington. No doubt, thirtieth-century human beings will wonder about the reason for our colossal sports stadia, magnificent golf courses, or lavish indoor tennis courts.

For the devotees of the ancient astronaut thesis, the pyramids represent the work of beings from outer space. Either they did the construction or taught men and women on Earth to do the job. According to one fanciful tale reported in W. Kingsland's *The Great Pyramid in Fact and Theory*, magic rods made the huge stones move through the air to their desired resting places.

Pyramids purportedly have a message of significance. Andrew Tomas offers many illuminating measurements for the Great Pyramid at Giza.[19] He claims the angle between the sides and the plane of the foundations is 51 degrees and 51 or 52 seconds; the perimeter of the base divided by the double height yields 3.1419, or pi; the height is the astronomical distance to the sun reduced one thousand million times, with an error of one percent; the number of pyramid-length units in one base is equal to the number of days in a year.

The senior lecturer in computer studies at the University of New South Wales, Australia, Phillip Grouse, has shown some of the errors in these numbers.[20] The area of the base of the pyramid of Cheops divided by twice its height, in feet, does not yield the value of pi; even if the cubit unit of measurement is used, the result is not 3.14.

Manipulation of numbers is, however, harmless compared to the more avid contentions of the pyramidologists who envision powerful forces emanating from the configuration and, in some cases, profit handsomely from the gullibility of the public. At first, the legend arose that anyone who enters the tombs is destined to die—as all of us are. During the 1970s, paper and plastic pyramid shapes in a variety of sizes were sold to sharpen dull razor blades, dehydrate eggs and meat, and improve the state of mind. Actress Gloria Swanson confessed to sleeping with a pyramid under her bed because it made every cell in her body tingle; actor James Coburn liked to meditate in his pyramid tent.

Alan and Sally Landsburg report on the commercial exploitation of pyramidology in Europe and the United States.[21] The history of the pyramidology cult is succinctly described by Martin Gardner.[22]

2

The Evidence from Archaeology

In *Chariots of the Gods?*, as well as in the film and the American television spectacular identically titled, much is attributed to the Nazca lines in Peru; they are presented as an airstrip fashioned by the extraterrestrials. However, the producer of the television show later wrote the book *In Search of Ancient Mysteries*, in collaboration with his wife, and presented an alternative interpretation. In 1941, Dr. Paul Kosok of Long Island University discovered their astronomical significance. At sunset on 22 June, the day of the winter solstice in the southern hemisphere, a line pointed exactly to the sun on the horizon. He checked out other lines and decided that the geometric figures in the desert could be "the world's largest astronomy book."[1]

Early peoples may have been deficient in written materials, but they read the book of the sky with great ability. Astronomical stone circles have been found in many places; Britain has about 150 of various sizes and dimensions, with the one at Stonehenge being best known. Astronomer Gerald Hawkins claims that the place was built over a period of centuries and was an astronomical observatory.[2] Alexander Thom in his *Megalithic Lunar Observatories* shows similar cases. In the western United States, a smaller stone circle, the Medicine Wheel in the Big Horn Mountains of northern Wyoming, has also been interpreted as an astronomical tool, built by Indians during the eighteenth century. Of course, the extraterrestrial enthusiasts can claim that the knowledge to build the structures was imparted by ancient astronauts.

Stonehenge

The thesis that intelligent visitors to the Earth are responsible for the accomplishments becomes soiled in the light of the large number of errors, faulty interpretations, and misinterpretations in the books of those who advance the idea. Theologian Clifford Wilson analyzed von Däniken's first book in this respect.[3]

Wilson cites the dogmatic statement about the huge colossi of the Olmecs of Mexico: they would never be shown in a museum because they were virtually immobile. The giant heads are found in museums, such as the Metropolitan Museum of Art in New York. Von Däniken contends that an island in the Nile River has an elephant shape when seen from aloft and is so named after the animal. But the Greeks called it "Elephantine," and the Greek word *elephantinos* has nothing to do with elephants. Again, the five-strand necklace of green jade in the burial pyramid of Tukal in Guatemala is not a miracle from China, because according to the experts on the Mayas, jade was found and used in this territory.

The ancient astronaut theme finds its evidence largely in the interpretation of artifacts, and an abundance of these readings are actually misreadings or at least open to other interpretations. Von Däniken sees a picture and calls it a portrait of a helmeted spaceman, while an expert on Aztec culture reports a Toltec soldier wearing a headdress and protective breast plate—the so-called communication equipment being a spear thrower. Andrew Tomas claims that in 1959 a shoe print was found on sandstone, millions of years old.[4] The reader is overwhelmed with "the Soviet-Chinese paleontological expedition led by Dr. Chow Ming Chen, which made the discovery, could offer no explanation of this strange find" in the Gobi desert. There is no evidence presented to show that the imprint is as old as the rock or that the imprint could be other than a shoe. Archaeological and anthropological artifacts have been misinterpreted by the best minds. An eminent physiologist of the nineteenth century, Rudolf Virchow, was shown the skull of a Neanderthal, and he viewed it as that of a sickly human with a bashed-in head.

One of the favorite citations of the ancient astronaut group is that of a skull, thousands of years old, having a neat round hole. The details vary from author to author: some say that the skull is that of a primitive man; others claim that it belongs to a bison; but the conclusion is always the same: only a bullet could have made such a hole, and bullets could only have been brought by extraterrestrials. Peter Kolosimo writes:

> These are the opinions of the curator of the Moscow museum, Professor Constantin Flerov. If he is asked who could have gone hunting bison with a prehistoric Siberian rifle, Flerov shrugs his shoulders and smiles. His colleagues are less careful than he is and do not hesitate to say: "Only one explanation is possible—the one linking it with the landing on Earth, at various times, of explorers from space very long ago."[5]

The opinion quoted is indeed as strange, if not stranger, than the so-called hole. There is no alternative interpretation offered in an area where fraud, deception, misrepresentation, and outright fabrication are more prevalent than in any other science. At the beginning of the century, for example, Dawson, in England, skillfully put together two pieces of different-age bones and had the anthropology community stymied for a couple of decades. His *Eoanthropus erectus* was a monstrous put-on for reasons no one has yet deciphered. At the time of Isaac Newton, a German professor was mercilessly led astray in his avid collection of fossils having imprinted letters, including his own initials: his colleagues had tricked him by planting the spurious articles.[6] So, too, the so-called bullet hole could be a fraud; could have been made in recent times; could be other than what is purported. Stoneley and Lawton have reported the bullet hole "to have been caused by a type of marine shellfish capable of boring regular-shaped holes of this kind, and the skull is now known to have been immersed in the sea at one period."[7]

A claim made by Jacques Bergier is readily disproved.[8] In 1962, a steel object weighing about twenty-two pounds fell in Manitowoc, Wisconsin, and is known to be a part of a Soviet artificial satellite. However, Bergier writes: "The object was taken to the Smithsonian, which announced that it was a manufactured object, then fell into silence. Today the object in question has disappeared under the dust of the Museum."

It is chiefly von Däniken who uses drawings on rocks as evidence for his theme. He sees spaceships and astronauts where others may not, just as tourists visualize rock formations to be kissing camels, a Model T Ford, or their favorite teacher. Von Däniken has some science-trained sympathizers in his camp of interpretation. Heinrich Gosswiler, a Swiss technical-drawings engineer, told the *National Enquirer* about an ancient cave painting in California: "Unbelievable as it seems, the drawing appears to be a blueprint for artificially inducing the elements of life."[9] Professor William Orme-Johnson of the University of Wisconsin's Enzyme Institute agrees: "It appears to show how subjecting certain substances, like formaldehyde and cyanose solution, to hightension electrical discharges will produce amino acids and small molecules. The painting certainly appears to show some of the pioneering experiments of the mid-1950s which attempted to create the basics of life."[10] Earlier, many including the tutored in science were excited about a "Martian god" in a space suit drawn on cliffs overlooking the Sahara. It turned out to be Tassili frescoes of an ordinary person in a ritual mask and costume.[11]

Those persuaded to visualize what von Däniken purports rock drawings to be should question some other attributes. For drawings made by extraterrestrials with great technological know-how, or by earth beings instructed by ancient astronauts, the diagrams are extraordinarily crude. An artist of the early Renaissance, not yet cited by anyone as being an extraterrestrial, has drawn spaceships that can be more clearly delineated. Hieronymus Bosch (1460–1526), in *The Temptation of Saint Anthony the Priest*, depicts a flying saucer with outstretched wings, "manned by a creature with an egg-shaped body and a head obscured by a hat resembling an inverted lamp shape."[12] Bosch depicted fantastic themes, and perhaps the ancients, too, had artists with imagination as well as those interested in producing replicas. The pre-Inca city of Tiahuanaco, pillaged by the Spanish conquistadores and perhaps the Incas before them, does have drawings that can be made out to be web-footed, four-fingered beings. Is this any more a picture of reality than Bosch's flying winged fish? Inca legends do have a goddess with infirmities in hands and feet, but Christian legends can also support the fantasy in *The Temptation of Saint Anthony the Priest.*

The ancient astronaut fraternity has not argued with artists. Almost unanimously, while clothing themselves in the fabric of science, they have attacked archaeologists. Von Däniken has been a leader in the assault. In his first book he claims: "Scholars make things very easy for themselves. They stick a couple of old potsherds together, search for one or two adjacent cultures, stick a label on the restored find and—by presto!—once again everything fits splendidly into the approved pattern of thought."[13] In his third book published in the United States, describing a tunnel system in Ecuador and Peru, he writes: "I'd like to see the archaeologist with nerve to tell me that this work was done with hand-axes!"[14] Dealing with carbon-14 tests, a widely used dating technique, he announces: "These are not proofs to be taken seriously, they are tricks to bluff us when scholars have nothing else to rely on."[15] At the First World Conference on Ancient Astronauts, April 1974, in a Chicago suburb, he told a newspaper reporter: "We have no scientific proof of these visits because too many scientists on this wonderful, blue planet are still sleeping."[16] Jacques Bergier contributes: "An archaeologist, even one of the present day, looking at a drawing of the LEM vehicle poised on the moon, would conclude unhestitatingly that it was an insect."[17] He also reports: "It is true that if we obtain a date for it very much prior to that of any known civilization, the archaeologists will protest vigorously, since such a fact would not jibe with their theories."[18] Peter Kolosimo uses a fiction story to mock the archaeologists: "The scholars would simply have glanced absent-mindedly at the heap of coins and then shrugged their shoulders and gone away muttering something like 'utterly impossible,' 'stupid,' or 'a childish prank.' "[19] W. Raymond Drake claims that "archaeologists seek man's origin in the mind, we search with shining eyes for our home in the stars."[20] Erich von Däniken and the others include paleontologists in their contumely. He complains: "Paleontologists generally know too little of man's record. And what they have never heard of they are not interested in. Myths and old books are rejected as fairy tales. They are terribly sure of themselves."[21]

The scientists respond, but seldom with the flamboyance of the ancient astronaut writers and with much less distribution in the mass media. McGuire Gibson, head of the University of Chicago's project in Iraq said:

> The most repulsive thing about von Däniken's thesis is that it presents man in a rather meager light, insisting that he has done virtually nothing on his own without outside help. If he or anybody else could prove some of the things he is saying we would jump on them so fast it would be incredible. Do you know what a premium it is for an archaeologist to find the oldest this or the biggest that? It's not as if archaeologists haven't read von Däniken's books, because we have. In fact we keep a copy of *Chariots* at our diggings for light reading, along with the *Wizard of Oz*. But people like von Däniken have not read our books.[22]

Most of the scientist critics complain about von Däniken's mistakes and misinterpretations. Professor Basil Hennessy, formerly director of the British School of Archaeology, Jerusalem, and a visiting professor of Near Eastern Archaeology at the University of Sydney, Australia, wrote that von Däniken had a "fascinating collection of unsupported, undigested, disconnected and often inaccurate claims."[23]

Timothy Ferris, the *Playboy* interviewer for the magazine's candid conversation with von Däniken, reported that an archaeologist who had studied von Däniken's work called him a liar.[24] His fabrication tendencies were affirmed by two psychiatrists. Examined when nineteen years old, he was said to display a "tendency to lie." A court-appointed psychiatrist described him as a liar and a criminal psychopath, just before serving a year in prison for embezzlement, fraud, and forgery.

Exobiologist and astronomer Carl Sagan of Cornell University offers another defect. He told the *Playboy* interviewer that von Däniken was ignorant of archaeology, had little understanding, and saw omnipresent evidence for extraterrestrial intelligence.[25]

Von Däniken very seldom has a direct confrontation with an archaeologist or scientist. In December 1973 he had a debate on Toronto, Canada, television with Dr. Ruth Tringham, a Harvard University anthropologist. Although she won, he captured a good percentage of the votes of the audience. Perhaps fewer would have been swayed had more of his egregious errors, mistakes, or lies been presented. A poll conducted by *Popular Archaeology* showed that forty-six percent of the respondents thought the ancient astronaut theme to be absurd, ten percent thought it was silly, and none found it convincing, although one percent called it reasonable.

In *The Gold of the Gods* he described caves in Ecuador where he allegedly saw artifacts of an advanced civilization —such as furniture made of plastic. Yet the South American adventurer who was cited in the volume, Juan Moricz, protests that he never took von Däniken into any such cave. His German publisher financed an expedition to investigate. A leading archaeologist familiar with Ecuador tried for six weeks to locate the place; none of the archaeologists in the country knew about the caves. Von Däniken admitted to the *Playboy* interviewer that he had not told the truth about the location of the place.[26]

Von Däniken's film, entitled *Chariots of the Gods?*, shows an allegedly prehistoric cave painting in Uzbekistan, depicting an astronaut and a flying saucer. An inquiry to the Soviet scientist who publicized the painting brought the reply that the picture was a modern one, not prehistoric.[27]

Von Däniken purportedly uses material from the Gilgamesh Epic. Noel Weeks, a lecturer in ancient history at the University of Sydney, Australia, whose work involved the translation of cuneiform texts including the Epic of Gilgamesh, wrote: "These events . . . [in] von Däniken's account of the Gilgamesh Epic *are not in the Gilgamesh Epic*."[28]

Professor Barend A. van Nooten, professor of Sanskrit at the University of California, Berkeley, told a May 1974 symposium at the university that passages in *Chariots of the Gods?* are not in the original Sanskrit texts of the Indian epics, *Ramayana* and *Mahabharata*. Von Däniken did not document his references and he does not read Sanskrit.

Could it be that the entire scientific establishment—particularly the archaeologists—are the liars, misinterpreters, and frauds, while von Däniken and cohorts are courageously attempting to sway public opinion? After all, scientists have been known to exaggerate, make mistakes, and be dogmatic. But the chances for this turn around are practically nonexistent. Von Däniken and friends filter their material only through the eyes of the public; archaeologists scrutinize each other's work, and every new discovery is thoroughly analyzed before being accepted into the body of knowledge.

3

Is It Religion?

Those who promote the concept of extraterrestrial visitations
have an abiding faith in the idea, although each has varia-
tions on the theme. They range in their choice and interpretation of their scriptures—from fundamentalists taking
words at face value to those liberal with exegesis. They
can be divided as well into followers of Western and Eastern
works; churchgoers and independent worshippers; evangelists and self-contained.

The promoters vary, too, in wealth. Similar to the riches
accumulated by contemporary religion peddlers Reverend
Sun Moon, Reverend Ike Hensley and his Universal Life
Church, or an earlier one, Father Divine, some have made a
fortune with the extraterrestrial idea. One advocate not so
economically fortunate claimed that von Däniken had hit
the cosmic jackpot.

The Western sect has its origins in the pioneering attempts of Immanuel Velikovsky. A physician and psychoan-
alyst trained in Europe, he became interested in the Exodus
of the Israelites from ancient Egypt. He wondered whether
a cosmic cataclysm had happened at the time and began
examining many ancient documents and records. Velikovsky
decided that old writings in many different parts of the

Earth did indeed describe such an event, and he published his idea.[1] He argued that around 1500 B.C. a gigantic amount of material came out of Jupiter, sideswiped the Earth and Mars several times, and finally became the planet Venus. With the help of this catastrophe, he accounted for such events as the biblical ten plagues in Egypt, the parting of the Red Sea, and the rain of manna from heaven. The reaction of scientists was typified by the letter that astronomer Dean B. McLaughlin at the University of Michigan wrote 20 May 1950 to the book publishers of Velikovsky's *Worlds in Collision*. He complained: "In my quarter century of experience as a research astronomer and teacher of astronomy, I thought I had seen just about every form of 'crackpotology' but the publication of *Worlds in Collision* by a nationally famous and erstwhile reputable publishing house is indeed 'something new under the sun.'"[2] Perhaps the expressions of outrage and attempted censorship by some spurred sales. By 1974, Velikovsky's books had sold more than one million copies, advocates had organized to defend him, and he finally had a session devoted to his ideas at the 1974 meeting of the American Association for the Advancement of Science.

The Velikovsky theme was one important factor spawning the new fundamentalism at the core of the ancient astronaut idea. The contention is that the Christian scriptures, taken at face value, describe extraterrestrial visitations.

Lesser-known religions are also conducive to support of the concept. In 1955, the Urantia Foundation, Chicago, published a two thousand-page "bible," *The Urantia Book*, describing seven superuniverses in the grand universe.

Generations ago practically all Western theologians had the fundamentalist viewpoint and even considered natural phenomena as signs from God. Historian Lynn White, Jr. illustrates the point in a story about sixth-century Pope Gregory the Great:

Gregory, not yet Pope, had seen English slaves in the Roman

slave markets, and decided to evangelize this pagan people. He received permission from the then Pope and started for England. On the evening of the second day out, while he was resting and reading, a locust—*locusta* in Latin—hopped on his book. He knew that God was speaking to him. The Latin word *locosta* means "stop"; he took this to be the meaning of the message and went no farther. The next day couriers from Rome reached him and summoned him back. The people of Rome had demanded that the Pope recall Gregory from what would have been a lifelong mission because they desperately needed his leadership.[3]

The bestiaries used during medieval times were in the same vein, with nature being used to illustrate some point of Christian teaching. The *Physiologus,* translated into Latin during the fifth century and into many others, such as Anglo-Saxon and Ethiopian, described the ant-lion, born of the ant and lion, with two natures, unable to eat either meat or seeds and which died in misery like every double-minded man who tried to follow both God and the devil.

The advancers of the extraterrestrial visitation hypothesis consider the Bible to have a record of the event. Only proper reading is necessary. Jean Sendy reads Genesis 1:14; God said, "Let there be lights in the vault of heaven to separate day from night and let them serve as signs both for festivals and for seasons and years. Let them also shine in the vault of heaven to give light on earth." But Sendy's interpretation is that "I have the impression that the text is referring to the making of maps of the sky. Maps to be used by the Celestials, since the sky seen from our solar system is quite different from the sky seen from another planetary position."[4] Von Däniken reads Genesis and contends that "the creation of the earth . . . is reported with absolute geological accuracy."[5] For him, Sodom and Gomorrah were destroyed by a nuclear explosion "according to plan."[6] Alan and Sally Landsburg list an assortment of items from the Old Testament, and each fits into the pattern:

There was Exodus's mighty words about "thunders and lightnings, and a thick cloud upon the mount . . . and Mount Sinai was altogether on a smoke, because the Lord descended upon it in fire." There was the good man Enoch who "walked with God" and was taken up by a "whirlwind" without having to die. There was Jonah, whose incredible sojourn in the belly of a whale might have been a trip in a submarine. There was the prophet Elijah, who heard a voice on Mount Horeb and departed from this earth riding in a "chariot of fire" which was "wrapped in a whirlwind."

There was Daniel's wondrous vision by a river of a being with "body flashing like the topaz, face like lightning, and eyes like lamps of fire." There were the various angels who came out of the sky and paid visits to Abraham, to Gideon, to Jacob, to Joshua, and to Lot.[7]

An engineer in the American space program, Josef F. Blumrich, confesses that he approached the ancient astronaut theme with a great deal of skepticism but was won over completely as he read and pondered about the experience of Ezekiel, described in the Bible; he experienced religious conversion. Blumrich views the words of Ezekiel as a report of a spaceship, one built with four helicopters.[8] Others may say that the visions of Ezekiel resulted from mushroom eating. Physicist Donald Menzel interprets the report of Ezekiel as *parhelia*, rings of light encircling the sun, caused by the passage of sunlight through a thin layer of ice crystals.[9]

The New Testament as a record of ancient astronaut activities is described by W. Raymond Drake in *Gods and Spacemen of the Ancient Past.* Another in the same vein is *The Bible and Flying Saucers*, by Barry H. Downing; he earned a bachelor's degree in physics and a Ph.D. in the relationship between science and religion. Among other points advanced in his book, he maintains that Jesus was taken into heaven by a spaceship after crucifixion; that such a vehicle led Moses and the Israelites out of Egypt and parted the Red Sea with its antigravitational beam. In 1975,

physicist Irwin Ginsbergh presented a similar theme in
First Man, Then Adam. He cited the Garden of Eden as a
spaceship, with Adam and Eve aboard, which crash landed
in the Mideast. Von Däniken, on the other hand, in *Miracles
of the Gods,* argues against the idea that Jesus was an
astronaut.

Clifford Wilson, the theologian mentioned in the prior
chapter, although a critic of von Däniken, presents himself
as a believer in the significance of unidentified flying objects, in his paperback book published in 1975, *UFOs and
Their Mission Impossible.* Evangelist Billy Graham, in his
1975 book, *Angels: God's Secret Agents,* claims that UFOs
are "astonishingly angel-like in some of their reported appearances." On the other hand, John Weldon and Zola
Levitt in *UFOs: What on Earth is Happening* present demons and fallen angels as responsible for UFOs; they are
signals for the Great Tribulation, a seven-year rule of the
Earth by the Antichrist. Then Jesus will return to establish
His Kingdom.

The new fundamentalists present personal gods, beings
from elsewhere who visited the Earth. In their church,
other theological perspectives would be outlawed. The new
fundamentalists have a church gathering. Eric Norman
describes the California Assembly of Bob Geyer and the
Church of Jesus the Saucerian.[10] England is the base for the
Aetherius Society, founded by George King, who views
himself as the "primary terrestrial channel"; he was contacted by the Masters from Outer Space in 1954. His society
offers pilgrimages, rituals, chants, and prayers. The Founding
Church of Scientology was incorporated in Washington, D.C.
in 1955, and an offshoot, the Church of the Final Judgement,
operates in England and Chicago—both sponsor doctrines
close to the ancient astronaut theme. However, a large percentage of the believers in extraterrestrial visitations would
probably view these groups as extremist rather than representative.

The typical ancient astronaut theorist would probably agree with the position of von Däniken:

> The religious legends of the pre-Inca peoples say that the stars were inhabited and that the "gods" came down to them from the constellation of the Pleiades. Sumerian, Assyrian, Babylonian and Egyptain cuneiform inscriptions constantly present the same picture: "gods" came down from the stars and went back to them; they traveled through the heavens in fireships or boats, possessed terrifying weapons, and promised immortality to individual men.[11]

Von Däniken denies the contention that the ancient astronaut theme is a religious doctrine. He so told the *Playboy* interviewer.[12]

The chaplain of Pulteney Grammar School, Adelaide, Australia, takes a different view. Reverend John Gent wrote:

> Von Däniken can see that primitive and ancient man was a myth-maker but absolutely and humourlessly he fails to see that he himself is trying to devise a new myth, a kind of scientific cargo-cult. Speaking of the "boundless stupidity" of which people are capable, he says: "There are groups who develop fanciful religious ideas from hitherto unexplained phenomena or build cranky philosophies of life from them."
>
> He would deny that he himself was developing "fanciful religious ideas" from inexplicable phenomena. And yet everyone of his readers is being invited by rhetorical question after rhetorical question to join such a group and to make just that sort of development.[13]

Most of the ancient astronaut writers extend their scriptures to include questionable and non-Western documents. Robert Charroux cites the Cabala, revered by a Jewish mystical sect. He also quotes from the Book of Enoch, off-limits for traditional theologians. Charroux reports that the book was retrieved from Abyssinia in 1772 and was copied from an original written in Hebrew, Chaldean, or Aramaic.

Chapter 7, verses 1 and 2 in the book, are purported to be support for extraterrestrial visitations:

> When the children of men had multiplied in those days, it happened that elegant and beautiful daughters were born to them. And when the angels, the children of the heavens, had seen them, they fell in love with them; they said to each other, "Let us choose women from the race of men and have children with them."[14]

The account is amazing in that the ancient astronauts are depicted as lecherous, and the Earth women are all beautiful. The visitors were technologically advanced, yet neither birth control nor biological engineering was within their purview.

In the Book of Enoch, Noah's father, Lamech, has doubts about the fidelity of his wife, Bat-Enosh. Lamech lamented, "I have brought into the world a child different from others; he is not like men, but resembles a child of the angels of heaven." Noah could be interpreted as an albino, but the ancient astronaut writers would have him be a product of the heavenly visitors.

Von Däniken bolsters his theme with the Egyptian Book of the Dead.[15] He and several others refer to the Book of Dzyan. He claims that the material is "a secret doctrine, was preserved for millenia in Tibetan crypts. The original text, of which nothing is known, not even whether it still exists, was copied from generation to generation and added to by initiates."[16] Actually the book has been promoted by an occult group whose principal figure is Mme. H.P. Blavatsky (1831–1891). Born in Russia as Helena Hahn, she married a seventy-year-old general when she was seventeen years old. She traveled in Egypt, Tibet, and India. In 1875, with her pupil, Colonel Olcott, she founded the Theosophical Society. She mentions the book for the first time in *The Secret Doctrine,* published in 1886. She is the high priestess of esoteric theosophy, yet the London Theosophical Society reported in 1908: "The Book of Dzyan is not in the possession of any European library, and was never heard of by European scholarship."[17] W. Raymond Drake reports that the document was "written in the sacerdotal language of Senzar."[18]

Drake has other fanciful inventions. He has space stories and space heroes in the major centers of the Orient.[19] He attributes spaceships, hydrogen bombs, and Earth visits to the extraterrestrial beings.

Drake refers to the Hindu epic *The Mahabarata*, a book more than twice the length of the Bible and seven times as large as Homer's *Iliad* and *Odyssey* combined. A very poor English translation exists, and a new one is being prepared by a Dutch-born professor of Sanskrit at the University of Chicago. The epic relates to India's past but also contains such myths as a mountain cohabiting with a river, embryos carried in the womb for one hundred years, and talking animals and plants. Here, too, women have been impregnated by the gods.

Drake presents the scripture of the ancient astronaut worshippers in greater detail than do the others, and at the same time presents much less of what could be considered evidence. He tells the Indian stories about an eagle with an extra head on its abdomen, lightning flashing from its beak and thunder from its flapping wings. Drake views this description as that of a spaceship.[20]

Von Däniken's reconstruction of past events has less theology. He begins with a battle between intelligent beings. The losers escape to the Earth and build a tunnel system in order to avoid the enemy. Finally, the losers emerge and build intelligence on Earth.[21] The account by George Adamski is similar. He would have the Venus people migrate to Earth about ten thousand years ago and oppose the Martians who taught man on Earth the game of war. Indeed, beings from several solar planets came to Earth and squabbled.[22] Jean Sendy contends that the Bible and other sacred books support the theme that Celestials arrived on Earth about twenty-three thousand years ago, stayed a while, and finally left.[23] Brinsley Le Poer Trench presents a different story. He claims that two types of men were created. Liberal and quick to learn, Galactic man was created in the Golden Age. Adam II was the product on the planet Mars of Jehovah, a captain among lesser gods. Adam II came to disobey Jehovah and he took them to Earth; the supreme god did not want Adam II destroyed.[24] In a later published book, Trench gives a more complete view of Galactic man:

> They were created perfect in spirit form or, if you like, as pure divine energy . . . these Sons of God were also capable of divine creation, which was their natural heritage, being who they were, the Sons of their Father. So, they followed in their Father's footsteps, as he would have wished, and used their inherent creative abilities. They, in turn, created universes![25]

Drake's stories could be the basis for several denominations, similar to the two American Baptists who founded three churches; for example, he believes in the lost continent of Atlantis as well as Lemuria or Mu, which was a large territory in the Pacific Ocean and which sank several million years ago. For him, the first people on Mu were bisexual giants. He appears to agree with von Däniken in one respect: "Greek legends suggest that after interplanetary war the Titans and Cyclops, a most ancient stellar race, retreated underground."[26] He cites the subterranean civilization of Agharta, "believed by some students to be the source of Flying Saucers."

Drake is also a prophet of doom. He writes: "Some sensitives allege that very soon our Earth will suffer immense convulsion, the Poles will become displaced and our civilization destroyed."[27] Ordinary religion offers an alternative in "repent and be saved," but the ancient astronaut theorists have thus far not proposed a way to avoid catastrophe.

All the ancient astronaut promulgators are apparently fundamentalist in their approach to their scriptures. However, every word is an accurate description only when it suits them. They also indulge in wide and varied interpretation of the writings in order to maintain their stand. When the *Playboy* interviewer challenged von Däniken on this issue, he replied that every theologian sifts data, accepting only what they like.[28]

Theology implies faith, and von Däniken has expressed his. He said, "I am convinced my visitors have deposited proof. Someplace, they have left us information that they came for this and this reason. We will learn they left in this and this direction; they were traveling at such and such a speed; and that they will be back at such and such a date."[29]

Professor Kees Bolle, speaking at a University of California seminar, Berkeley, May 1974, viewed the ancient astronaut theme as a pseudoreligion, or at least akin to religion, with its concept of a Golden Age long ago. Instead of the nostalgia of the good old days, depicted in some conventional literature, von Däniken and the other promoters of the idea cite the Golden Age in outer space, with a technologically proficient being. If so, perhaps the appeal of a religion based on a Golden Age has its greatest popularity during times of social, economic, and political stress.

4

Is It Fiction?

One of the oddest true stories of all times is the twentieth-century abdication of Edward VIII of England for the love of Wallis Warfield Simpson, a twice-married, fortyish American, definitely not the most beautiful woman in the Western world. In 1974, the bizarre accounts of the kidnaping, alleged conversion and disappearance of newspaper heiress Patricia Hearst was ample evidence that truth is stranger than fiction; it seems unlikely that anyone could compose a more fantastic tale. However, the books dealing with extraterrestrial visitations are laden with stories staggering the imagination. Perhaps the authors learned from the H.G. Wells novel *War of the Worlds,* or from the 1938 radio broadcast drama by Orson Welles and his Mercury Theatre group, describing a Martian invasion centering about New Jersey. The motion picture *2001: A Space Odyssey,* completed in 1968, postulated a slab placed under the moon's surface about four million years ago. Twentieth-century fiction, along with other entertainment forms, began to have shock value.

Jacques Bergier appears to have a special fondness for outlandish material.[1] He relates several stories about appearances and disappearances of human beings directly or indirectly connected to visits on Earth of space beings.

Bergier's chapter eight of *Extra-Terrestrial Visitations from Prehistoric Times to the Present* is devoted to Kaspar Hauser, an individual who came to Nuremberg, Germany, May 1828. His vision was perfect, his skin white, and the soles of his feet were soft. He did not walk properly and was void of many social amenities. This is sufficient evidence for Bergier to zero in and label Hauser an extraterrestrial. The usual explanation that he was an abandoned child who grew to manhood away from civilized centers is not considered. Hauser was murdered little more than five years after he surfaced, and serious investigation of his origins were halted.

In presenting Hauser as an extraterrestrial, Bergier was following the lead of a godfather of the occult, Charles Fort. In his book *Lo!*, the story of Kaspar Hauser is not only given in detail, but there also appears the statement: "One might feel oneself driven to the alternative of believing him to be a citizen of another planet, transferred by some miracle to our own."[2]

Bergier attempts to bolster his case with several other accounts. In 1909, a boy in Wales left no trace of his whereabouts while on his way to a water well. A few pages later the explanatory supposition Bergier would favor is apparently available in the story about the Archbishop Agobard of Lyons (779–840), who had three people stoned because they claimed to have gone to Matagonia in an airship. If the reader objects to ancient history and the involvement of only one or two human beings, there is another tale a few pages ahead about the Danish training ship *Kobenhaven* disappearing in 1928 after leaving Montevideo in South America. Chapter fifteen in John A. Keel's *Our Haunted Planet* has an array of similar disappearance stories.

Extraordinary organisms in human form are cited by Bergier. Following the Kaspar Hauser chapter is one about the green children. On one afternoon in August 1887, two children with green skin came out of a cave near the Spanish village of Banjos. The boy died a month afterwards; the girl lived five more years, became a domestic servant, and learned a few words of Spanish. She said that she had come from a land without sun. In the same chapter, twelve-foot monsters seen near Flatwoods, Virginia, in 1952 and the twelve-foot humanoids allegedly seen in Oregon in 1963 are mentioned. At the other extreme is the black dwarf captured in the cellar of a German monastery in 1138. The dwarf was released; he returned to his tunnel and no one succeeded in following him.

An antidote to Bergier's claims is available in a metropolitan hospital visit. Observing infants at birth over a long period of time reveals a large variety of disfigured bodies, including six-fingered, organ deficient, and multicolored. A stay in the same city will also show how simple disappearance without leaving a trace can be for almost anybody. Police annals of missing persons are large, and periodically a solved case shows that the apparently lost individual has adopted a new identity—or was even encased in cement. Newspapers have also printed accounts of children abused and held in chains for years, or military draft evaders living in garrets and away from society for long periods of time.

Physicist Philip Morrison has shown how a large piece of equipment shipped on a known train route and presumably securely tied could disappear for decades. Despite intensive searching, the motor stator and rotor assembly could not be found. Twenty years later at the draining of a swamp, the object was found in the mud. Apparently when the train had rounded a curve, the ties had become unfastened, and the material had rolled into the swamp.[3]

Each of Bergier's tales can be checked and investigated, even though the scholarly apparatus of references and citations is missing from the book. However, a small amount of judicious reasoning can also cast reflection upon the stories. Thus, the three French medieval people punished for their

flying experience could have been using a delusionary nar-
cotic producing effects of what is known as a "trip".[4] Likewise, the black dwarf could have been in the imagination of
drunken prelates.

Other fascinating entertainment is in Bergier's book. He
quotes from the apocryphal Book of Enoch. Bergier's Enoch
receives a visit from very tall men whose "faces shone like
the sun, and their eyes were like two burning lamps. And
fire shot forth from their lips. . . . Their feet were purple."
Enoch visits seven different worlds, seeing oddities, and
he also meets the creator. Enoch's confession that his trip
lasted only a few days, but that he returned to find the
passage of centuries, does have the ring of time dilation
according to the theory of relativity. However, just a few
years before World War II, science-fiction writer Gerald
Heard minutely described what came to be in 1943 the
atomic-energy city of Oak Ridge, Tennessee. Whoever wrote
Enoch, as well as Gerald Heard, had a good imagination.

Bergier's stories indicate a variety of extraterrestrial
visitors. In his chapter on those from the Middle Ages, he
has the father of renowned mathematician and physician
Jerome Cardan receive seven men in silken garments, purple
undergarments, and glittering boots; one had a dark red face.
For the nineteenth century, Bergier conjures up Springheel
Jack, luminous at night and armed with a device giving a
blue flame and an odor of ozone.

Beginning on page 528 of his book *The Occult,* Colin
Wilson describes the strange visitors to Jack Schwarz, well-
known in psychic circles as being able to withstand a nail
through his palm. Wilson relates how in 1958 a tall Arab
appeared on his knees before Schwarz and said, "You are
my master." The individual reappeared on other occasions
in other disguises and claimed, "We can appear in any shape
or form we desire. . . . We come from a tribe of people
who crash-landed in a rocket ship on earth thousands of
years ago."

Several others in the ancient astronaut fraternity offer

fantastic tales of visitors from elsewhere. Eric Norman contends that Oliver Cromwell (1599–1656), before his greatest victory on the evening of 3 September 1651, when he led his armies against Charles II, had a conversation in the woods with a monk-robed figure who offered Cromwell a role of parchment. Norman cites a witness who heard Cromwell shout something about broken words, seven years, "I don't like it," and "accept the terms."[5]

Andrew Tomas claims the source of a tale to be an Arab writer named Muterdi. A man was trapped and presumed lost during the exploration of the Khufu Pyramid, but he reappeared later and spoke in a strange language.[6] The reader is undoubtedly expected to associate the phenomenon with the pyramid's alleged strange powers.

Tomas reports a more recent incredibility, embellishing a newspaper story in the Los Angeles Times, 22 May 1932. Residents in the Mount Shasta area told of seeing white-robed, tall, white men, with closely cropped hair and a band across their foreheads. They gave merchants gold nuggets in excess of the value of goods purchased. Strange cattle were reportedly theirs. Tomas adds, "To add to the enigma, rocket-like airships have been observed over Mount Shasta territory. They were wingless and noiseless, sometimes diving into the Pacific Ocean to continue out on the sea as vessels or submarines."

Peter Kolosimo invokes the weird and eerie, relating the case of John Spencer in Mongolia in 1920, in a room with a green light and having coffins. One cadaver was dressed in silver, and instead of a head there was a ball of silver with round holes where the eyes should have been, small openings instead of a nose, and no mouth.[7]

Ivan T. Sanderson reports that a French movie crew in Morocco in 1928 saw a large number of fourteenth-century ships sail out of the Atlantic Ocean and into the sky, disappearing over the Sahara Desert. One man said he saw a Turkish emblem on one ship; others identified pennants belonging to Portugal and Spain.[8]

Both Sanderson[9] and Eric Norman[10] as well as von Däniken[11] are intrigued by the stone disks found in caves in the mountains on the border of China and Tibet. Allegedly, a Chinese archaeologist interprets them to be a record of a spaceship landing twelve thousand years ago; analysis in the Soviet Union showed a high percentage of the element cobalt. Each disk resembles a phonograph record, having a center hole and a double groove of spirals from the center to the circumference. In between the grooves are hieroglyphics, and one of them translates to: "The Dropas came down from the clouds in their gliders. Our men, women and children hid in the caves ten times before sunrise. When at last they understood the sign language of the Dropas, they realized the newcomers had peaceful intentions." The authors indicate that the Peking Academy of Pre-History found the material too shocking and, for a time, banned its publication. A greater oddity, it appears, is that reputable scientists in the USSR and China, interested in extraterrestrial communication, have ignored the disks.

Soon after Erich von Däniken told the story in *Gods From Outer Space*, Dr. Kwang-chuh Chang of Yale University, a specialist in Asian archaeology, familiar with every dig conducted in China about the time the stones were reputedly found, confessed his ignorance of the matter. Moreover, he claimed there was no Chinese archaeologist named Chi Pu Pei nor a Peking professor named Tsum Um Nui—both mentioned by von Däniken. Dr. Chang indicated the names sounded to him like words concocted by a Westerner.

Von Däniken's account of the stone-disk affair is even more outrageous. He would have us believe that beings from outer space were stranded on Earth and "had been hunted down and killed" by Earth people. How could an advanced intelligence not be able to cope with early man on Earth?

The stone disks are in a category with the account in the infamous *Book of Dzyan*, wherein there is a description of the great "lance that traveled on a beam of light," allegedly a rocket and its jet of flame.

The storytellers need neither logic nor observation for support, building tales with the help of unfettered imagination. Robert Charroux clothes one of his as a letter received from a reader. The letter offers the information that the extraterrestrials are Baavians from a planet belonging to Proxima Centauri, four and one-third light years away. Their secret base is on an atoll of the Maldive Islands in the Indian Ocean. The Baavians first went to Mars, were intimate with Martian women, and twelve thousand years ago when Mars deteriorated they came to Earth.[12] In *One Hundred Thousand Years of Man's Unknown History*, Charroux argues that our "superior ancestors" had a nuclear disaster, and man has since climbed back "up the scale of evolution."

Peter Kolosimo credits one of his tales to Beltrán Garcia, a Spaniard who wishes to revive sun worship and claims to be a descendant of an Inca princess and a Spanish conquistador. On page 195 of Kolosimo's *Timeless Earth*, published in 1973, he purports that five million years ago a spaceship landed on the Island of the Sun in Lake Titicaca and "from this there alighted a female creature resembling a woman from her breasts to her feet but with conical head, huge ears and webbed hands with four fingers."

Those who read such fantasy may become imbued enough to act. Von Däniken claims he received about twenty letters from United States residents claiming they were extraterrestrials and inviting him to a rendezvous.[13]

Andrew Tomas points to a Chinese alchemist who drank an antigravity liquid and became airborne; moreover, his dogs and poultry did the same. Tomas cautions: "Let us not ridicule these curious tales because many fantasies of the Orient have become real through modern science."[14] Another Oriental story he relates is about the inhabitants of vast underground shelters "flooded with a peculiar light which affords growth to plants and gives life to this lost tribe of pre-historic mankind."[15]

Charitable critics can classify the fanciful accounts as science fiction. If that is their category, they are a strange variety, based upon discounted or unacceptable scientific themes. W. Raymond Drake readily admits that Hans Hoerbiger's World-Ice theory is "somewhat discredited by official science" but nonetheless proceeds to suppose the development of civilizations on Earth during three periods of Earth history between three successive moons.[16] A Viennese mining engineer, Hoerbiger, published his idea in a large book in 1913. A cult soon developed about the thesis, and the Nazis in Germany took it up. After Hoerbiger died in 1931, an English student of mythology, Hans Schindler Bellamy, promoted the idea. According to him, our present moon, captured about 13,500 years ago, has a thick coat of ice, as does Mercury, Venus, and Mars; moreover, the Milky Way is a ring of gigantic slabs of ice.[17]

It is evidently no fiction for Drake that in 1952 a Dr. G.H. Williamson purported to have had radio and telepathic communication with a planet called Hatonn in the Andromeda galaxy.[18] After all, Superman, popular on American radio during the 1940s, came from the planet Krypton.

Peter Kolosimo's science fiction is also unusual. He reports a stone appearing to be a snake and emitting a man from its mouth.[19] He suspects a monolith on the moon, a little closer to science fiction today. However, the London Daily Telegraph, 10 July 1970, reported the claim of Argosy magazine that Russian and American spacecraft had photographed "two groups of objects arranged in definite geometric patterns and appear to have been placed there by intelligent beings." Brinsley Le Poer Trench cites the Australian Flying Saucer Review for July 1970, with the opinion of a Soviet space engineer that the objects on the moon were in a pattern known in ancient Egypt as an "abaka"; the centers of the spires of the lunar abaka being arranged in exactly the manner as the apices of the three great pyramids.[20] All of this may be viewed in the light of some activity at the Houston Space Center and Cape Kennedy. During the Apollo moon explorations of the early 1970s, some scientific workers even speculated about a secret door on our natural satellite leading to wonders inside.

Scientific and engineering workers, among others, do read and write science fiction. The celebrated astronomer Fred Hoyle has several stories to his credit; his *Black Cloud* postulates strange life-forms drifting among the stars. John R. Pierce, using the pseudonym J.J. Coupling, is the author of eleven science-fiction books.[21] It has been suggested that one reason why some aerospace engineers have been attracted to the ancient astronaut theory is that they are imbued with science fiction. Professor Michael A. Arbib of the University of Massachusetts at Amherst, speculating about extraterrestrial intelligence, wrote: "But whether that intelligence is anything at all like that of humans will be a very open question, and our attempts at an answer will at times be closer to science fiction than to science."[22]

Another reason for the entrance of the ancient astronaut theme into the chambers of reality is the Shaver experience in science fiction. Early in 1945, *Amazing Stories* published a piece entitled "I Remember Lemuria!" The author was a Pennsylvania welder named Richard B. Shaver, but it was widely believed that the editor of the magazine did somewhat more than edit. According to the tale, people on Earth are descended from a race of giants who went to distant planets about twelve thousand years ago. They left behind some, called abandoned, who went into underground caverns together with teleportation gadgets, rock-piercing long-distance rays, space rockets, and rejuvenating apparatus. These cave dwellers degenerated into dwarfs who harassed human beings. According to *Life* magazine, the story was taken seriously by many and created "the most celebrated rumpus that rocked the science fiction world."

Many authors took up the theme. In 1969, Robert Ernst Dickhoff's book *Behold—The Venus Geruda* postulated giant winged humanoids in space, seeking out human beings on Earth as food. In *Agharta,* he claims that Martians who went back about eighty thousand years ago made a network of tunnels on Earth. "They left behind a secret metropolis called Rainbow City, which still serves as an interplanetary space port."[23]

The myth of Atlantis is not considered fiction by some because Plato described the continent. Lemuria was first suggested as an Indian Ocean abode to account for the geographical distribution of the lemur. Theosophy and anthroposophy took up both places as a home for some of the "root races" of man and provided fiction writers with much substance; for example, the Toltec on Atlantis were copper-colored, tall, drank hot blood of animals, and had airships. Lemurians, according to the founder of anthroposophy, could lift enormous weights and communicated via telepathy.[24] Mu was supposed to have had seven great cities and many overseas colonies. Allegedly started about 150,000 years ago, the centers were subject to a continental disaster about 12,000 B.C. and were destroyed.

5

Is It Science?

Humility is never a characteristic of the ancient astronaut theorists. They promote their idea with the fervor of evangelists, and positive, dogmatic statements abound.

R.L. Dione writes in the Introduction to his book: "I present the material in this book, not as a theory based on some sort of mystical revelation or metaphysical hunch but rather as a collection of pure, hard facts—facts based on statistical evidence."[1] Jean Sendy claims "the hypothesis that forms the structure of this book is founded on portions of the Bible."[2] Erich von Däniken charges: "Because it has long been the custom to hammer into us as schoolchildren the presumptuous idea that man is the lord of creation, it is obviously a revolutionary and unpleasant thought that many thousands of years ago there were unknown intelligences who were superior to the lord of creation, but however disagreeable it is, we had better get used to it."[3]

If humility is a characteristic of scientists, the ancient astronaut theorists are not to be so classified. Yet they do want to be viewed as objective investigators.

Von Däniken and the others, despite their attacks on archaeologists, associate themselves with well-known scientists, almost like the entertainment-business luminaries who have momentary, fleeting, or indirect contact with someone, then claim them as best friends or refer to them in terms of endearment. Von Däniken lists exobiologists apparently in league with him.[4] He also gives names of those in sympathy with the idea of prehistoric visits by extraterrestrials.[5] Jean Sendy dedicates a book to the memory of Giordano Bruno.[6] Jacques Bergier in his prologue writes: "I would rather compare myself to those eccentric characters who, before the appearance of *Origin of Species* published 'bizarre' books—from which Darwin learned."[7]

The authors seemingly take pleasure in presenting themselves as injured parties, not fully appreciated by our society but whose time will come. They cite many ideas and inventions of the past, ridiculed when first stated but applauded later. Of course, they ignore the many more conceptions forever not accepted, and associate themselves with the pioneers who were responsible for electric lights, airplanes, and space travel. Andrew Tomas devotes many pages to myths later found true. Among his appealing accounts: Bouillard demonstrating Edison's phonograph before the Paris Academy of Sciences in 1878 and being accused of ventriloquism and fraud; chemist Antoine Lavoisier, father of modern chemistry according to the French and many others, claiming that meteorites could not fall from the sky; Heinrich Schliemann believing Homer's *Iliad* and finding the ancient city of Troy.[8] Tomas also clothes himself with a modern space pioneer by writing in the Preface: "Professor Herman Oberth has confessed to the author that Jules Vernes's *From the Earth to the Moon* prompted him to convert a romance into rocketry formulae." The publisher did not exploit such contents; the classification on the spine of the book is *occult*.

The mention of scientists in the books with the ancient astronaut theme could persuade an unsophisticated reader that the idea was accepted in scholarly circles. However, scientists reject the works. An editorial in *Science*, signed by the editor, Philip H. Abelson, remarked: "The readers of earlier works generally understood that they were scanning fictional material, but the new books seek to create the impression of scholarship and verity. *Chariots of the Gods?* does this in several ways. It has a bibliography. In an Introduction it acknowledges help from personnel of the National Aeronautics and Space Administration, including Werner von Braun."[9] Two scientists reporting in the *New Scientist*, 1 April 1976, that the manna that fed the children of Israel in the desert may have come from a single-cell protein plant, concluded: "It is tempting to speculate, with many other contemporary writers, that the Earth was visited by space people some 3000 years ago and to go on to suggest that it was these visitors who provided the machine. This raises as many problems as it solves and we would prefer not to propound such a hypothesis today."

The books with the ancient astronaut theme are unscientific in at least three ways: first, they reject as false some almost completely scientific principles; second, they misrepresent science and scientists, let alone attempting to associate themselves with the discipline; lastly, their procedure at arriving at results bears little resemblance to established, scientific ones.

The theory of organic evolution has been attacked by religious fundamentalists, and even today they pressure governments and schools to restrict or modify its dissemination. The ancient astronaut fraternity opposes the conception on similar grounds—conflict with their scriptures. Robert Charroux writes: "It is unlikely that man descends from the apes."[10] He also reports: "The laws of evolution can scarcely teach us anything about our destiny, for the fact is that evolution has no scientific rigor and cannot be proved. . . . No real links connecting man and ape have ever been found."[11] Immanuel Velikovsky said: "Most controversial is the evolutionary question. I have done a great deal of work on

Darwin and can say with some assurance that Darwin . . . did not derive his theory from nature but rather superimposed a certain philosophical world-view on nature and then spent twenty years trying to gather facts to make it stick."[12] The sixth chapter in Richard E. Mooney's *Colony Earth* is entitled "The Evolution Hoax."

The fact is that organic evolution is an ancient idea. Darwin, as well as Alfred Russel Wallace, collected much data in its support. Both men almost at the same time developed a mechanism—natural selection—for the concept to be found, for example, in the writings of Aristotle. Charles Darwin expressed his debt to this great philosopher.

Erich von Däniken specifies his objection and concommitantly indicates a misunderstanding of the very slow nature of evolutionary changes. He claims:

> There is really nothing in the theory of evolution to explain the mighty leap by which *homo sapiens* set himself apart from his family of homonids. All we hear is that the brain suddenly became efficient, acquired technical know-how, was capable of observing the heavens and establishing communication in social communities. In terms of the history of evolution this leap from animalistic being to *homo sapiens* took place over night.[13]

He evidently views the animal world as void of reasoning and other abilities.

Fundamentalist-religion opponents of organic evolution have been known to use the same tactics. They, too, cite a false issue and proceed to demolish it; and one of their most-used arguments is what might be called the "missing link." During the last century, Bishop Samuel Wilberforce did as much in his debate with Thomas Henry Huxley before the British Association for the Advancement of Science, when he asked whether Huxley's monkey ancestry was from his mother's or father's side. Evolutionists have clearly indicated that the slow development from anthropoid ape to *homo sapiens* is through many species, some found as fossils, and there is not one single man-ape or ape-man.

Max Flindt and Otto O. Binder present a somewhat different perspective than those who attack organic evolution.[14] They proceed to question early fossil records of unspecialized organisms and describe man's unique physiology and anatomy. They would like some so-called missing links to explain away man's hairlessness yet head of hair, copious tears, flexible hand and fingertips, full-color vision, and lack of a penis bone. Other physiology they find unique include the low healing rate of human skin, the lack of tooth gaps, slow swallowing, facial mobility, and large penis. These features and man's enormous brain and cerebral cortex pushes them to the belief that Earth people are a colony of creatures established by ancient astronauts. The latter must have been an advanced species, but Flindt and Binder do not speculate about genetic engineering such as cloning.

During the last century Alfred Russel Wallace was a leader, along with Charles Darwin, in postulating natural selection as a mechanism for organic evolution. Wallace, however, a firm believer in phrenology and spiritualism, could not come to apply natural selection to man. He, prior to Flindt and Binder, was impressed with man's consciousness, intellectual and moral nature, hairlessness, erect posture, delicate, expressive features, smooth skin, and beauty.

The ancient astronauts described by all the authors appear to possess ability in the physical sciences and engineering but are not too sophisticated in biological techniques. Knowledge of the molecule of heredity, DNA, for example, should have given the "star people" ability to fashion mice and men at will. Why bother with the long process of organic evolution? However, Alan and Sally Landsburg in their second paperback with the ancient astronaut theme, *The Outer Space Connection*, suggest cloning and genetic engineering.

Flindt and Binder do not misrepresent—they misinterpret. Chapter seven of Jacques Bergier's *Extra-Terrestrial Visitations from Pre-Historic Times to the Present* is a good, albeit extreme, example of how the ancient astronaut adherents misrepresent science and scientists.[15] Bergier purports that Sir Henry Cavendish (1731–1810), an outstanding English scientist after whom the Cavendish Laboratory at Cambridge University is named, was an extraterrestrial or in communication with one. Bergier's evidence—the word is abused in this instance—is that an attempt was made to conceal his birth. Bergier is amazed, too, that Cavendish was admitted to the Royal Society at the age of twenty-nine. Cavendish shunned society, was a very wealthy man, and accomplished much in scientific work. Bergier twists these simple facts: he was asocial because he did not want human beings to see him; the source of his immense wealth is not known; he had outside help; and for one aspect of his work he "received instructions not to make this . . . public." Brinsley Le Poer Trench, when describing Comte de St. Germain, a man who knew many languages, music, alchemy, and never appeared to age, wrote, "St. Germain could well have been a spaceman living among us," but at least added immediately: "This, I realize is wild speculation."[16]

In speculating about Cavendish, Bergier also gives a unique view of the Age of Reason—the time during the days of Sir Henry Cavendish when the Western world became enthusiastic about science and human reason. According to Bergier: "In the middle of the eighteenth century, an information source X was introduced. The event took place about 1730. Some one (one or more) stopped limiting himself (or themselves) to obtaining information and making inquiries and instead brought important information to Europe, notably with regard to physics and chemistry."[17]

Bergier makes history bend to his needs by having Kepler regard his story of travel to the moon, *Somnium*, "his fundamental work."[18] Bergier misrepresents the English-American historian of science, Derek deSolla Price, by quoting him with "It's rather frightening" in regard to the antikythera machine retrieved from the sea and credited to the early Greeks.[19]

The misrepresentation of contemporary, rather than earlier, scientists is more widespread. The favorite ones for ancient astronaut adherents are Carl Sagan, an astronomer at Cornell University; I.S. Shklovskii, an astrophysicist in the USSR; Freeman Dyson, a physicist at the Institute for Advanced Study in Princeton; and Frank Drake, a U.S. astronomer. Bergier ties Drake to the ancient astronaut concept: "It also is not out of the question that manuscripts and objects left by extraterrestrial beings await us in the area's caves. Frank Drake, a U.S. scientist believes that such caves were marked with radioactive isotopes so that they would only be discovered by an advanced civilization."[20] Bergier gives the same treatment to Gerald Feinberg, a physicist at Columbia University who has speculated about material with a speed exceeding that of light—tachyons. In Bergier's book, Feinberg is given the opinion that signals from extraterrestrial intelligences are via tachyons.[21] Fortunately, Bergier does not name those who, in private, cite pulsars as artificial, rather than rotating, neutron stars.[22]

Alan and Sally Landsburg quote from Will Durant's *Story of Civilization* about primitive cultures and how they survived, to lend support, where none is present, for their contention. They write: "This gives some scholarly respectability to the idea . . . that visits and/or seeding at long intervals by an advanced race from outside our solar system may be the cause of man's sudden great leaps forward followed by long periods of stagnation or regression."[23] Nowhere does Will Durant mention civilizations from outside our solar system.

Misrepresentation need not occur because the ancient astronaut proposers have their own band of luminaries to support the theme. They are not authorities; they are not accepted in established science; but their remarks add spice to the main dish. W. Raymond Drake cites Hans Hoerbiger:

the Austrian cosmologist (1860–1931) theorized that Space must be permeated with hydrogen and water vapour. The Universe materialized the conflict between cosmic fire and ice, symbolizing the eternal struggle between Good and Evil. This controversial theory postulates that the Space around the Sun is filled with ice-particles; the inner planets, Mercury and Venus, are covered with ice. . . . Earth has apparently had several moons. In Tertiary times, date unknown, the predecessor of our present Luna spiralled close to the Earth, attracting the world's waters into a girdle-tide obliging the population to flee from the flooded coastal plains and climb higher and higher up the Mountains, establishing lofty asylums in "Andinia," Mexico, Abyssinia, Tibet, New Guinea and elsewhere.[24]

Charles Hoy Fort (1874–1931) is one often-cited figure supporting the ancient astronaut theme. He believed that mankind was owned by higher intelligences who periodically visited the Earth to check their property. Fort was a writer whose passion was collecting information not in agreement with science, such as the apparent fall of frogs from the sky. His writer friends formed the Fortean Society and issued the *Fortean Society* magazine.

Misrepresentation of scientists is never an acceptable procedure, but if it were the only transgression, perhaps forgiveness or benign neglect would be appropriate. The ancient astronaut theorists, however, indulge in other unacceptable techniques, with a very much used one being the transformation of supposition into fact. Alan and Sally Landsburg illustrate this many times throughout their book.

The Landsburgs asked Dr. Leslie Orgel, Salk Institute, LaJolla, California, a distinquished scientist: "Do you believe that directed panspermia caused the origin of life on Earth?" "No", he said, "it's just an interesting possibility." And then he left.[25] The reader is never advised how the word *possibility* is intended. A few pages later, however, the Landsburgs repeat: "One planet might have been spraying all the nearer solar systems with spore-carrying rockets, say once every hundred thousand years or so, as a sort of gift to posterity or just as an experiment, to see what would happen."[26] They add fuel to the supposition of extraterrestrial intelligence colonizing other solar systems by describing a star with no hydrogen, a contracting core, and an expanding atmosphere; intelligent life must get away. "Therefore, the space probes are sent out to locate habitable planets."[27]

When the Landsburgs describe the gold earrings worn by Incas of the highest rank, supposition may soon become fact to the reader who notes the questions: "Could the earpieces have been receivers? Or could they have been connected to wires implanted in their brains?" The next two paragraphs lead the gullible forward. One describes current brain implantations, and the next reports: "If the Inca knew some way to enhance their health or intelligence by wiring themselves, the gold earpieces would conceal the wiring."[28]

The procedure becomes intoxicating; it is repeated ad infinitum. The reader is asked to remember what the Apollo astronauts did on the moon.

> First they established a safe base, and cautiously tested survival possibilities. Next they scouted the terrain close by, gradually widening their sweeps. They studied possibilities of creating shelter from materials at hand. They looked around to see if the supplies they brought with them could be augmented from local sources. They made geological surveys. They pinpointed landmarks as navigation aids for longer journeys later.

The reader who goes along is then asked to believe: "Let's

say something like this happened about 10,000 B.C. or earlier."[29]

On page 7 of their second book, *The Outer Space Connection,* the Landsburgs continue to offer supposition in the guise of fact, reporting:

> I wanted to locate an area described by John Stephens, a nineteenth-century explorer who was responsible for the rediscovery of lost Mayan cities. In *Incidents of Travel in Central America,* he had reported finding an artifact he described as a smooth, black, glasslike surface which Mayan priests consulted when making crucial decisions. Stephens made no other mention of the object. I felt that his perspective in examining it was limited by the perception of nineteenth-century man. In light of today's knowledge, a black, glasslike surface, used for communication, might well have been a television receiver.

The Landsburgs fit the Sumerians into the extraterrestrial ideology by assuming they came forth suddenly "with the first written language, sophisticated mathematics, a knowledge of physics, chemistry, and medicine." Ignoring material about the emergence of civilization, the Landsburgs suppose that cataclysmic events on Earth destroyed many of the extraterrestrial colonies, and that the Sumerians represent "fragments of a highly advanced culture struggling for survival."[30]

The authors of books about ancient astronauts usually intertwine supposition and fact so that readers do not separate the two and accept the package offered without criticism. Erich von Däniken concocts a scenario to explain why snakes are so omnipresent in creation stories. According to him, primitive peoples saw the jet vehicles of ancient astronauts in action and compared them to a "feathered fire-breathing serpent."[31] Bergier points to the famous iron pillar of Delhi after a discussion of data storage and retrieval, and ponders, "I still have to ask if it is a giant recorder." He is impressed with various "artifacts" that have turned up in old coal deposits, and he glides easily to his thesis through rhetoric and supposition. "These objects most often belong to private collectors, who refuse to entrust them to scientists and, lacking a study actually demonstrating the contrary, we may . . . admit the possibility that these angled objects have come from outside and were not manufactured on earth."[32]

Louis Pauwels and Jacques Bergier in *The Morning of the Magicians* use the same technique of presenting supposition as fact. An artifact in the Near East is presented as an ancient chemical battery, although it may have been many other possibilities. They write: "If a German engineer, Wilhelm König, had not paid a chance visit to the Museum at Baghdad, it might never have been discovered that some flat stones found in Iraq, and classified as such, were in reality electric batteries, that had been in use two thousand years before Galvani."[33]

The Fire Came By, written by a professor of theatre arts and a journalist, serialized in some newspapers in 1976, interprets the Siberian Tungus phenomenon of 1908, usually attributed to a huge meteorite, as a nuclear spaceship explosion. Evidence for a nuclear holocaust is supposed, not presented.

The least persuasive procedures are used by the promoters of the extraterrestrial-visitation hypothesis. First is analogy, which may be a pedagogic tool for introductory material but is hardly a means to establish a conclusion. When Galileo saw the first four moons of Jupiter, the opponents to the heliocentric conception could rightfully ignore this analogy to the sun and the planets. Thus, the famous iron pillar of Delhi is analogous to a "giant recorder," and only to those imbued with the first scenes of Stanley Kubrick's science-fiction motion picture, *2001*.

Another low-grade procedure employed, albeit many thinkers, notable and not, embrace the same technique, is the construction of correlations. It is possible to postulate a connection between any two items: the length of your hair can be related to the breakfast of teenagers on the Marshall Islands; the temperature of air in March can be closely associated with the amount of sexual intercourse in July by chimpanzees in Tanganyika. Those who tie sunspots to a variety of phenomena engage in the same kind of structuring. The number and size of sunspots have been correlated with baldness, wars, economic depressions, stock-market prices, and the occurrence of rheumatism. What is to prevent, then, connecting huge, fashioned rocks found in various parts of the Earth to visits by ancient astronauts? The procedure is hardly different than the cutting open of chicken hearts to determine whether your next few hours will be productive, or climbing atop Mt. Olympus to read your future in the cloud formations.

Statisticians can protest a blanket indictment of correlations, in that high, positive ones uncover new truths. Sunspots are related to radio reception as well as precipitation on Earth, although the exact manner of connection is unknown; likewise, cigarette consumption is connected to the incidence of lung cancer. However, the ancient astronaut promoters offer no statistics with their correlations.

Induction and deduction are other thought processes improperly used by the ancient astronaut theorists. They rush to conclusions upon the basis of one or two, often questionable, instances. They not only amplify the significance of an artifact of their choice but also throw other really "harmless" pieces into the category of evidence. Bergier collects his array of addities, such as Kaspar Hauser and Henry Cavendish, and expects an inevitable conclusion. More likely, Bergier has already swallowed the main thesis and indiscriminately applies it, expecting others to do the same.

Another characteristic of the scientific enterprise violated by the ancient astronaut theorists is replication, or verification by competent personnel. When a scientist reports an experiment or observation, it is not fully accepted until confirmed by another scientist. Violators suffer. During the early twentieth century, expert biologist Paul Kammerer was accused of using India Ink to paint his specimens showing inheritance of acquired characteristics; no one else could duplicate his feat. In 1974, a similar experience befell physician researcher William T. Summerlin.[34] In the last quotation from Bergier's book, for example, readers are asked to accept his interpretation of artifacts never examined by anyone save private collectors.

Close to the procedure of replication is that of checking out facts—a necessity in proper copy-editing as well as in science. Yet, in *Chariots of the Gods?*, von Däniken describes the iron column in Delhi, India, as having resisted rust for thousands of years and being made of "an unknown alloy from antiquity." Investigation completed before he wrote his book showed the column does have rust and the composition is not unusual, but von Däniken did not bother to confirm his data.

The writers have a penchant for mysticism and secrecy—traits not acceptable in scientific circles. Robert Charroux has written several volumes pushing the theme of ancient visitations and, in one Foreword, proposes that "Robert Charroux is not the sole author of this book. . . . He has been instructed by exalted beings, some of whom may be unknown Masters of the World."[35]

Critics can suggest that the ancient astronaut writers do no more than the scientists in applying a theory. They attempt to fit the facts to the idea. An early twentieth-century English physicist said as much: "The object is to connect or coordinate apparently diverse phenomena and above all to suggest, stimulate and direct experiment." Von Däniken and his cohorts have done the first half, but nowhere in their activities is the second half of the statement. Indeed, von Däniken's comparable dictum, which he told the *Playboy* interviewer, would be a disaster to scientific accomplishment; von Däniken would simply ignore as misunderstanding whatever is not accounted for by a theory.[36]

Other scientists and analysts have indicated the work of a scientific theory, and in each case the ancient astronaut conception fits only partially. J.J. Thomson's student, the father of nuclear science, New Zealander Ernest Rutherford, said that the worth of a theory depends upon "the number of experimental facts it serves to correlate and upon its power of suggesting new lines of work." The theme of ancient astronauts has a very limited number of observational facts within its purview and has not spawned new observation nor experimentation.

Albert Einstein contended that a theory is "more impressive the greater the simplicity of its premises, the more different kinds of things it relates and the more extended is its area of applicability." The idea of extraterrestrial visitation does have very simple premises but does not score high for the other criteria.

A theory has acclaim in science if it permits the asking of questions that can be investigated; for example, James Clerk-Maxwell's electromagnetic theory of light opened several windows of the spectrum of radiant energy and is still a fund for the future. The periodic table of chemical elements has also yielded definitive questions leading to results. What new material has the ancient astronaut theme uncovered?

An acceptable theory in science is consistent with other established knowledge. When the ancient astronaut crowd enters the house of science, those who believe in organic evolution, among others, must leave, engage in continuous dispute, secrete themselves in a corner of the basement, or hope that others in the community will brand the newcomers as frauds.

It may be, as the ancient astronaut fraternity claims, that they have a revolutionary conception akin to the Copernican that jostled scientists over a few decades. In any event, their theory, if it is to be called scientific, must describe, explain, and predict in an efficient manner. Thus far, the descriptions have been misleading because observation and their theory have been presented in an intertwined fashion; explanation has been minimal because the events and artifacts encompassed by their view are just as puzzling; and prediction is nonexistent.

6

Is It Probable?

Any science-fiction story can be considered a possibility because the realm of what could be is indeed infinite. Thus, the immediate destruction of the Earth or the conversion of a stone to a human being are possible. That a band of typing chimpanzees will be able to compose the great works of mankind within a year is as possible as the dogs on Earth becoming the dominant species next month.

Probability, however, is a much more intelligent consideration. Probability means viewing factors in the light of the past and with the power of accumulated knowledge and wisdom. None of the events mentioned, therefore, are probable.

Many enthusiastic statements about ancient visitations cite a possibility more suitable to the science-fiction television show "Star Trek." For example, they would have an intelligence beyond that on Earth operate a device capable of forming a human being into a ball of inanimate matter, and vice versa. W. Raymond Drake writes: "In theory a cosmonaut could be transformed into photons modulated to his key-frequency beamed towards a distant star, then on arrival reconstituted to his normal body, a science fiction conception if ever achieved would still limit motion to the speed of light, leaving the problem of galactic travel yet unsolved."[1]

The alchemists, claim the supporters for instant transformation of the form of matter into the desired configuration, had goals reached only today. Why cannot men be changed into mice or rock and, most important, vice versa? This is a confusion of possibility with probability. The latter is not only what can be done now but also what may be accomplished in the near future.

How probable are ancient astronauts? The writers of books with the theme cite authorities in support. Eric Norman quotes Professor M.M. Agrest, who visited the Baalbeck platforms in Lebanon.[2] The professor, a student of ethnology, not archaeology, wrote that he was convinced it could be, not that it was a launching pad. The professor is also of the opinion that many of the events described in the Bible can be interpreted as astronaut visitations, and he cites the *Book of Enoch* to support these contentions.[3]

Norman also quotes Dr. Thomas Gold, professor of astronomy at Cornell University, about space travelers visiting Earth a billion years ago.[4] Gold states a possibility, not a probability. He does not state facts in support of an hypothesis, but simply a suggestion suitable for more investigation.

Norman gives the same treatment to Professor Carl Sagan. At the 1966 meeting of the American Astronautical Society, Sagan speculated, "Our tiny corner of the universe may have been visited thousands of times in the past few billions of years. At least one of these visits may have occurred in historical times."[5] The words are conditional and tentative, not definite and dogmatic.

The scientist who spoke after Carl Sagan at the 1974 American Association for the Advancement of Science discussion in San Francisco on "The Search for Extraterrestrial Life" gave him a unique reference. He said that Carl Sagan was always a tough act to follow, particularly since he has all kinds of things to talk about—some of which may be real and some imaginary; the more imaginary, the more fun.

Andrew Tomas turns the trick of authoritative quotation a different way. He cites Professor Frederick Soddy, an early Nobel Prize winner in chemistry, saying in 1909, "It may be an echo from one of many previous epochs in the unrecorded history of the world, of an age of men which have trod before the road we are treading today."[6] Soddy's words, a suggestion rather than a conclusion, support a cyclical view of history rather than the ancient astronaut theme.

Reputable science and scientists can be used for a more definite support of the extraterrestrial-visitation hypothesis. For one, the principle of the uniformity of nature indicates the existence of other solar systems, some like our own. The sun is an average star, and billions abound in the observable universe. Those suns could support a family of planets, and perhaps the ones similar to the Earth are widespread. The uncounted number of such planets is where intelligent life reigns.

The nature of the intelligences in the universe can be analyzed through the Gaussian distribution curve—the bell-shaped figure describing the distribution of any large number of items. Heights, weights, educational attainments, wealth and taxes paid within the United States, for example, as well as intelligences in the universe, show a large number of average, a small amount below the central tendency, and an equally small figure for above the average. If cynics dominate, and intelligence on Earth is said to be below normal, the more advanced life elsewhere must be far ahead of us. If we are humble and claim to be average, then only a small percentage top us. What if intelligence on Earth is among the best variety in the universe? Then the ancient astronaut idea falls.

A favorite ploy of the believers is to cite an "authority"

who views human life as being on the low side of the scale. Thus, Dr. James Harder, an engineer at the University of California, was asked, "Do UFOs pose any threat? Do we have reason to fear them?" He answered, "If you pick up a mouse in a laboratory situation, it's very frightening to the mouse. But it doesn't mean that you mean the mouse any harm."[7] Or Flindt and Binder report that Professor Nikolai Kardashev of the Soviet Space Institute believes a supercivilization exists in our galaxy.[8] Thomas Aylesworth, publicizing his 1975 book for young readers, *Who's Out There?*, said, "If an ant came up to you and said that it wanted to have a chat, you wouldn't step on it. You would be curious enough to try to have the conversation it requested." On page 67 of the 1975 book by Hynek and Vallee, *The Edge of Reality*, Hynek reports, "If you were a bushman in Australia, how many parts from a Boeing 747 would you be picking up?"

Logic and science can be used to probe mankind's intelligence vis a vis others in the universe. Starting in a belief in the big-bang theory for the development of the observable region, the investigator can form a tentative idea about relative thinking abilities of beings. The cosmic evolution, or big-bang hypothesis, is the thought that billions of years ago the universe was very concentrated, and a kind of explosion sent the material spreading apart. Evidence is twofold. Spectral patterns of the galaxies show a red shift, indicative of increasing distance; and the background radiation accompanying the big-bang was detected in 1965. Should the theory be correct, the age and development mechanisms of the galaxies would appear to be very similar. There would not be much in the way of galactic head starts.

The ancient astronaut theme is concerned with stars supporting solar systems, rather than whole galaxies with billions of stars. In any one galaxy, the stars vary in characteristics.

Early in this century, two astronomers, Henry Norris

Russell and Einar Hertzsprung, independently made a correlation apparently connecting in some fashion several star attributes. Known as the "Hertzsprung-Russell Diagram," the chart was at first an academic novelty. Later, the data was used to support a theory of how stars change with the passage of time. This concept of stellar evolution seemed successful when so-called youthful, middle-aged, and older stars were observed. Stars like our sun, and others capable of having planets with life, should therefore be around along with those past the sun's development stage. The latter would have planets with civilizations beyond our sun, technologically.

There are several flies in this ointment. For one, stars change from those like our sun to enormous cool ones called red giants, and in the process all the planets are absorbed. The chance for the greater development of intelligence is gone. Secondly, theories in science—astronomy is not an exception—are continually being modified—that of stellar evolution seems destined for change. Within a few decades, ideas about the manner in which stars develop may have an entirely different cast. Finally, and most important, about ninety percent of the stars in our galaxy are of the type called K and M, and these have characteristics not beneficial to life on planets. They have a short life-span, and extraordinary flares emanate from them—this radiation would probably be lethal to life on the accompanying planets.[9]

Another conservative scientific procedure can be used to judge the ancient astronaut idea. For example, to arrive at the conclusion that the Earth is spherical, is rotating and revolving, a sum of evidence is available. Adherents to the ancient astronaut conception have a paucity of material, such as a questionable artifact or a line on a desert.

In 1976, a book called *The Sirius Mystery*, by Robert Temple, an American living in England, postulated that the Dogon tribe living in the Republic of Mali are aware of a visit to Earth about five thousand years ago by a race of mermaidlike creatures from outer space. His evidence is the tribe's apparent knowledge (allegedly for generations) of the extreme density of the companion star to Sirius—the first white dwarf star discovered—and their veneration of it as shown in wood carvings, coins, paintings, and textile design. Facts and artifacts of civilization find their way through many channels; one could similarly claim that the Eskimos know industrial chemistry because DDT is found in Arctic snows. Likewise, Dean Jonathan Swift, as well as Voltaire, accurately described the moons of Mars long before they were discovered. Were the two authors in contact with extraterrestrial sources, or is it more likely that they had read outstanding pioneer astronomer Johann Kepler's argument, or some later version of it, that Mars should have two moons?

The complaint that Copernicus, Brahe, Kepler, and Galileo had minimal evidence for the heliocentric theory is correct. The phenomenon of stellar parallax—the shift in position with time of the near stars with respect to the far stars—a bulwark for the idea, was not detected until early in the nineteenth century. Again, current leaders in physics refer confidently to the graviton—the unit of gravitation—and the quark—the unit of elementary fundamental particles—as though they have a real existence; no one has detected the items. Are, therefore, von Däniken, Bergier, Sendy, and the others the unappreciated scientists of our day?

Large numbers of ideas in science are wrong when first proposed and remain wrong forever. It is more likely that the ancient astronaut conception is in this category.

A few friendly critics point out that some scientific literature would give credence to the ancient astronaut theory. William J. Kaufmann III, at one time director of the Griffith Planetarium in Los Angeles, writes:

The possibility that an astronaut could leave our universe,

travel along a timelike path all the way around the cylinder, and return to our own universe. By pointing his spacecraft in the appropriate direction, he can return to almost any point in space-time in our universe that he chooses. For example, he could come back to the earth a billion years ago or a billion years in the future.[10]

This is fiction. Likewise, imagination is the proclamation by a Soviet astrophysicist, N.S. Kardashev, that there are three general types of civilizations in the universe: one like our own; one capable of channeling the entire output of its parent star; and one with the ability to use the energy of its entire galaxy. Evidently, his Gaussian distribution curve for intelligence in the world has man in the lower depths. Another Russian, R.G. Podolny, at the 1971 conference on communication with extraterrestrial intelligence held in Soviet Armenia, cited recent Earth visitations. His evidence was an 1842 document about an event in Russia in 1663. He reported that "a sphere appeared about 4 meters in diameter; from the lower part two rays extended earthward and smoke poured from the sides of the vehicle."[11] Unlike the other participants in the conference, Podolny apparently had no academic connections and listed his affiliation as "Knowledge is Power—Moscow".

The 1971 conference was jointly sponsored by the Soviet and American Academies of Science. It was attended by Soviet and American scientists as well as by a few Britons. The conference concluded "that the chances of there being extraterrestrial communicative societies and our present technological ability to contact them were both sufficiently high that a serious search was warranted."[12]

Professor Ronald Bracewell of Stanford University, an electrical engineer, has suggested that probes be sent into space to monitor over a long period of time any intelligence coming to the Earth. According to British author John W. Macvey, an alien space probe has been in our solar system

for about thirteen thousand years.[13] His evidence was the
suggestion that trains of long-delayed echoes heard during
the 1920s could be interpreted as a message from an interstellar probe. Fuller investigation has given better natural
explanations.[14] Nonetheless, in his 1974 book, *In Search of
Ancient Gods*, Von Däniken presses for the extraterrestrial-
communication interpretation.

Extraterrestrial beings may be imagined who are not
communicative; indeed, one American scientist said he
would "hang up" if ever an extraterrestrial being contacted
him.

Occasionally a scientist is swayed by one so-called piece
of evidence or another. The reviewer of Jacques Bergier's
*Extra-Terrestrial Visitations from Ancient Times to the
Present* in the English journal *New Scientist* was so impressed with the Piri Reis map.

Admiral Piri Reis (1470–1554) of the Turkish Navy had
published a navigation atlas with the maps and had indicated
their origin: In a naval battle in 1501, the Turks took a
Spanish prisoner who had been with Columbus on three of
his voyages and had in his possession a set of maps showing
unusual features. There was supposedly a complete outline
of North and South America, Greenland, and Antarctica—
all unknown at the time. According to Andrew Tomas, a
Commander Larsen of the U.S. Navy said, "The Hy-
drographic Office of the Navy has verified an ancient chart—
it's called the Piri Reis map, that goes back more than
5,000 years. It is so accurate, only one thing could explain
it—a world-wide survey."[15] According to Jacques Bergier,
a Captain Arlington H. Mallery showed that the construction of the map had required knowledge of spherical trig-
onometry. More accolade was given by Professor Charles
H. Hapgood of Keene State College in Keene, New Hampshire.[16]

Clifford Wilson's criticism is that the map shows the
Amazon River twice, has nearly a thousand miles of coastline

missing from the east side of South America, and Antarctica is shown as a land mass joined directly to South America.[17] Another critic claims that Antarctica is not shown on the map.[18] Moreover, the author of *History of Cartography*, Leo Bagrow, where the map is also reproduced, reports that the map was found in the Topkapi Seraglio in 1935, and was not discovered in the eighteenth century.[19]

Another criticism assumes a forgery. In 1965, Yale University announced possession of a map that indicated that Lief Ericson had discovered America five hundred years before Columbus. That map, too, showed an accurate outline of Greenland. Analysis of the Vinland map by a Chicago firm specializing in small-particle analysis, Walter C. McCrone Associates, showed it to be a forgery dating from the 1920s. The ink contained a higher percentage of titanium used in inks of long ago; the yellowed lines were really yellow-browned ink; and the titanium mineral used—analase —was regularly shaped, indicating modern manufacture.

The Piri Reis map may not be a forgery. Does it mean support for the ancient astronaut theme? Perhaps the best guide is the opinion of Philip Morrison, a physicist at the Massachusetts Institute of Technology. Through a report in the British journal *Nature*,[20] he had, in collaboration with Guiseppe Cocconi, more or less stimulated interest in the topic of interstellar communication. He is book-review editor of the *Scientific American* and, reviewing Shklovskii and Sagan's *Intelligent Life in the Universe*, wrote: "Here is a body of literature whose ratio of results/papers is lower than any other." In 1974, he reported: "Now, the transport of material structures over galactic distances with a speed approximately that of light is probably beyond all accomplishment. In my opinion we shall never do it."[21] He also claims ". . . real interstellar travel, where people, intelligent machines, or whatever you like, go out to colonize. You travel in space as Magellan circumnavigated the world. I do not think this will ever happen."[22]

If man, now and in the future when presumably he will be more technologically superior, will not travel beyond our solar system, how can other beings do so? Even if man is considered to be at the bottom of the heap vis a vis intelligence, due time will give him much more capacity. Yet Morrison's opinion is for now and the future.

Scientific opinion is conditioned by the tradition of more rapid changes in accepted theory and the habit of adding the phrase "as far as is known" to every positive statement. The scientist may therefore appear tentative, halting, and uncertain as he outlines alternatives. But Morrison's statement is definitive, not hesitant.

Before space travel to the moon and sending probes into the solar system was a reality, scientists could be found who were very dogmatic about the impossibility of either feat. An engineer associated with the Naval Research Laboratory was featured in an educational film where he demonstrated how voyages to the moon "staggered the imagination." Could Morrison be making a comparable blunder?

Space travel is of two distinct and entirely different varieties: one, the accomplished was once a dream; the second—to the stars and galaxies—continues to be a dream because of the vast distances involved. The nearest star is reached by light from the sun in four and one-third years. Vehicles with even one-third or one-fourth the speed of light are not on the drawing boards, whereas the principle behind rockets for solar-system exploration has been known for generations.

The cautious person may desire every item uncovered by the ancient astronaut promoters to be disproved. Supporting a negative such as "these lines were not made by beings from elsewhere" or "these monuments were not constructed through the help of extraterrestrial intelligence" is a stupendous task. It should be sufficient that the major points are shown to be untenable.

The adherents will not easily give up the idea. Indeed,

new life will be breathed into the corpse with every facet of nature capable of being misinterpreted. Thus, in 1972, a team of French scientists found a sample of uranium in a mine at Oklo, the Republic of Gabon, Africa, with a very low percentage of U.235, the isotope used for atomic bombs and nuclear reactors.[23] Those who are attached to the ancient astronaut belief will say that long ago a civilization taught by spacemen used the material; scientists see a natural reactor as the answer.

Another approach seeking greater probability for the ancient astronaut theme views the present human dilemma in contrast to earlier generations when men would build with stone, use natural air-conditioning, and live well with nature. Therefore, a cynical cultural historian would have, for whatever reason, a stupid individual was introduced to the Earth long ago because the quality of life on Earth has steadily been declining.

The concept of extraterrestrial visitations long ago or even during historical times must be given a very low probability of occurrence because compelling evidence is lacking and the idea offers no reliable guide to the past, present, and future. The chances of the event are exceedingly slim. Another indication of its probability is the amount of money being spent in scientific research on the topic. Despite the outpouring of popular literature, funds for the investigation of ancient astronauts have been practically zero.

7

The Money Involved

When scientific research started in the Western world, its financial support was very limited. The only person who had an enviable situation was astronomer Tycho Brahe. The Danish king gave him an entire island plus necessary buildings and money for expenses; when that munificent deal ended, the king of Bohemia supported his efforts at a castle near Prague. Other scientists in western Europe either did experiments and observations during their own personal time, were associated with universities requiring a modicum of teaching and thus were free to do research, or had wealthy patrons for a few years. In France some men, members of the Royal Academy, were paid by the king, but this was unusual; most early investigators of nature, doing work requiring little expenditure of funds, were self-supporting.

When science became a professional activity, there was still a tradition of very little money for the work. Private industry gave funds for scientific research, but in the United States at the end of the last century the total available was about three million dollars.[1] During the late nineteenth century, public controversy ensued in both England and the

United States about the wisdom of financing scientific research. The Devonshire Commission in England gathered evidence for three years and finally reported in 1875 that science had become a foundation for general welfare and was worthy of support. In the United States the Allison Commission operating during President Grover Cleveland's administration wrestled with the problem of tax-supported scientific work. Financing the United States Geological Survey had little opposition, but a concept comparable to a national laboratory was heresy.

During World War I, scientists and engineers demonstrated their importance for the war effort, but their usefulness was not taken as commonplace—else there would not be examples among all the major combatants of outstanding young researchers being killed in combat; in England, for example, H.G. Moseley was struck by a Turkish bullet.

Although private donors gave increasing amounts of money to laboratories and observatories, public science did not become a viable idea until World War II. The atomic-bomb project, costing about two billion dollars, marshalled scientific resources under government auspices, and the result was convincing to both citizen and legislator. Government-sponsored scientific work became established, grew, and is now by far the largest single contributor to the scientific enterprise.

As is the case for practically all public funds, the amount of money spent on ancient astronaut or related-topic research is a matter of public record. It can be examined by anyone willing to search for and read the pertinent documents. The work in this instance is not arduous because the funds used have been extremely tiny and are all in related topics, rather than directly in the realm.

In 1960, Frank Drake, now at Cornell University, used the eighty-five-foot radio telescope at the National Radio Astronomy Observatory, Green Bank, West Virginia, for two weeks to study radio signals from two close sunlike stars—

Epsilon Eridani and Tau Ceti. This search for extraterrestrial intelligence, called Project Ozma, did not produce results. The funds involved were comparatively insignificant, but it is the first expenditure of money for a subject close to the idea of ancient astronauts. The negative conclusion of his study did not impede Drake from writing in his popular book two years later: "At this very minute, with almost absolute certainty, radio waves sent forth by other intelligent civilizations are falling on the earth."[2]

The next money spent came from the National Aeronautics and Space Administration. Beginning in 1966 it sponsored eight faculty fellowship programs each summer in cooperation with the American Society for Engineering Education. Four sites close to a NASA facility and university were selected, and a research-oriented as well as an engineering-design-oriented project was installed. In 1971 the NASA–ASEE Summer Faculty Fellowship Program in Engineering Systems Design was conducted jointly by Stanford University and the Ames Research Center of NASA; it was a design study of a system for detecting extraterrestrial intelligent life. Called Project Cyclops, its objective was "to assess what would be required in hardware, manpower, time and funding to mount a realistic effort, using present (or near-term future) state-of-the-art techniques, aimed at detecting the existence of extraterrestrial (extrasolar system) intelligent life."[3] The rationale of the venture could have been assigned to education because the participants heard lectures, held discussions, and presumably learned. Money was also used in the preparation of a 243-page, 8½-by-11-inch paper-covered final report available to the public. One conclusion printed was that the cost of a detecting system would be six to ten billion dollars, to be spent over a period of ten to fifteen years. Conclusions numbered twelve and thirteen are germane:

12. The search will almost certainly take years, perhaps decades and possibly centuries. To undertake so enduring a

program requires not only that the search be highly automated, it requires a long term funding commitment. This in turn requires faith. Faith that the quest is worth the effort, faith that man will survive to reap the benefits of success, and faith that other races are and have been, equally curious and determined to expand their horizons. We are almost certainly not the first intelligent species to undertake the search. The first races to do so undoubtedly followed their listening phase with long transmission epochs, and so have later races to enter the search. Their perseverance will be our greatest asset in our beginning listening phase.

13. The search for extraterrestrial intelligent life is a legitimate scientific undertaking and should be included as a part of a comprehensive and balanced space program. We believe that the exploration of the solar system was and is a proper initial step in the space program but should not be considered its only ultimate goal. The quest for other intelligent life fires the popular imagination and might receive support from those critics who now question the value of landings on "dead" planets and moons.[4]

The conclusions were written at a time when NASA funding was being cut and scientific research in general was being threatened with loss of funds. The conclusions did not state so, but previous suggestions by scientists for expensive projects have borne fruit when installed.

Just before World War II, refugee scientists badgered the president of the United States to alert him about atomic energy. The Manhattan District, Corps of Engineers, became the custodian of the atom-bomb project. Again, when the government became very liberal in awarding money, a group of geologists formed the American Miscellaneous Society and, partly in jest, proposed drilling a deep hole in the Earth. This developed into the costly, now defunct, Moho project, where new drilling techniques were uncovered.

In 1975 the Soviet Academy of Sciences outlined a plan in *Icarus* claiming that "efforts to detect extraterrestrial civilizations should proceed smoothly and systematically, and should extend over a long period of time." They urged that every star within one hundred light years of the sun be investigated, different galaxies be examined, and an all-sky survey be made. On the other hand, Professor Sir Martin Ryle, Britain's Astronomer Royal, called for a stop to unilateral attempts to communicate with extraterrestrial life. Late in 1976 he told the editors of *New Scientist* that he does not believe that there is a "hope in hell" of communicating with such life.

The results of a search for extraterrestrial intelligence can be negative for a long time and, as a consequence, will have almost no chance of being funded by governments. The United States and the Union of Soviet Socialist Republics are the only countries with sufficient funds to initiate, collectively or otherwise, the research. But in both countries, costly research must be practical to be supported.

If the United States alone is to be asked for several billion dollars, its past record spells an answer of "no, thank you." Not a single project has had appropriations a decade in advance; funds have been distributed yearly or so, with budget requests and cross-examination by legislators on the same time cycle. Moreover, scientific research in the United States has a spotty record of financing. The enthusiasm for science soon after World War II was repeated immediately after Sputnik in 1957, but periods of abrupt decline have been just as numerous. In 1967, Viet Nam War demands cut off many scientific projects, and in 1972 the Nixon Administration did not see fit to give added support to the laboratory; the office of scientific advisor to the president was abolished.

Prior to 1972, the United States did spend more money on a topic related to ancient astronauts. Dr. G. Verschuur, at the National Radio Astronomy Observatory, listened for signals from several stars. He used the 140-foot and 300-foot telescopes at Green Bank, West Virginia, and tried to detect intelligent signals from Barnard's star, Wolf 359,

Luyten 726–8, Lalande 21185, Ross 154, Ross 248, Epsilon Eridani, 61 Cygni, Tau Ceti, and 70 Ophiuchi—stars giving an indication of having solar systems. In the USSR, a group under V.S. Troitskii at Gorki University, and another at Shternberg Institute, have done similar radio-telescope work; the Troitskii project had a small radio telescope and studied two hundred stars. More recently, in the United States, Dr. Ben Zukerman of the University of Maryland and Dr. Pat Palmer of the University of Chicago used the radio telescopes at the National Radio Astronomy Observatory for an examination of about five hundred stars, and the results were negative. In 1974 and 1975, Bridle and Feldman at the Algonquin Radio Observatory in Canada, as well as Drake and Sagan at Arecibo in Puerto Rico, tried to detect signals from extraterrestrial intelligence. Herbert Wischnia, president of Sonitrol, Worcester, Massachusetts, intends to look for flashes of ultraviolet laser light, using the ultraviolet telescope aboard *Copernicus*, the orbiting Astronomical Observatory satellite. In November 1974, data was obtained for Epsilon Eridani during fourteen of the satellite's orbits around the Earth; and in 1975, Tau Ceti and Epsilon Indi were scanned. However, all the searchers, whether by radio or laser, believe a long time span, rather than sporadic moments, would be more realistic for their activities.

A relatively tiny amount of money was spent on the design and fabrication of a plaque for the Pioneer X satellite going into outer space. The plaque is the work of Dr. and Mrs. Carl Sagan and Frank Drake. It will last for eons and will presumably signal our presence to other intelligences capable of intercepting it. After the Arecibo Ionospheric Radio Observatory in Puerto Rico was resurfaced, radio signals capable of being deciphered by intelligent beings were sent in the direction of Messier 13. However, a minimum of about twenty-five thousand years must elapse before receivers on Earth will be able to detect a response.

Most moneys spent by the United States on topics related

to ancient astronauts were for symposia, lectures, and
meetings concerned with detecting intelligence beyond our
solar system. A series of lectures was organized at the NASA
Ames Research Center in the summer of 1970, and a commercial publication resulted.[5] A symposium was held at
Boston University, 20 November 1972; a Boston University
Astronomer, now a dean at American University, Washington, D.C., Richard Berendzen, edited the manuscript, and
NASA offers the small book for sale.[6] The conference held in
Soviet Armenia in 1971 resulted in a commercial publication,
albeit offered by a university press.[7]

The public—American and otherwise—have in another
sense generously supported the ancient astronaut promoters
—not for the purpose of doing scientific research, but for
the composition of entertaining books, magazine articles, and
television programs. Books purporting ancient visitations of
Earth by non-Earth beings have proved to be popular and
profitable. *Erich von Däniken Tours International* is now
available for those who wish to visit the places cited in his
books.

II
Introduction to UFOs

Public interest in unidentified flying objects, UFOs, began when Kenneth Arnold, a United States businessman, cited in 1947 an array of moving sky objects not fitting any known category. Using his description, the media tacked the title of "flying saucers" to the phenomenon.

The United States government became concerned enough to initiate "Project Saucer" at the end of 1947; the official code name of the investigation was first Project Sign, then Project Grudge, and finally in early 1952, Project Bluebook. The astronomer in charge of Project Bluebook for about twenty years, J. Allen Hynek, nearing retirement from academic responsibilities in 1974, started his own organized effort to understand what was reported. Public pressure was responsible for the study at the University of Colorado under physicist Edward U. Condon, but publication in 1969 of the results of the Condon investigation was not satisfying to those enthusiastic about the extraterrestrial-vehicle hypothesis to explain UFOs.

The believers in ancient astronauts and the UFO people aligned to an extraterrestrial concept are in two camps, without much intermingling. They are friendly but not friends. Not many follow theologian Clifford Wilson, a critic of von Däniken but an avid follower of UFOs. More subscribe to the words of Robert K.G. Temple on page 213 of *The Sirius Mystery*:

"I do not believe that spaceships from extraterrestrial civilizations are flitting through the skies at this moment, I am not

85

a 'flying saucer' enthusiast. I do not believe that spacecraft would behave in the erratic fashion in which UFOs behave. It makes no sense to me that spacecraft would fly idly around making spectacles of themselves—and ambiguous spectacles at that—for years on end."

Flying-saucer organizations were the first on the scene, and consequently they are more numerous. The ones with the most membership are the Aerial Phenomena Research Organization (APRO) and the National Investigation Committee for Aerial Phenomena (NICAP); the latter had twelve thousand members in 1968. However, Contact, chaired by Brinsley Le Poer Trench, calls itself "the largest UFO movement in the world." The Midwest UFO Network, headquartered in Quincy, Illinois, is a relative newcomer but growing. Each group today has a tiny staff.

On 23–25 June 1967, the Congress of Scientific Ufologists was held at the Commodore Hotel in New York City. Present among the enthusiasts was Edward Condon, as well as "Vivanus," a young brunette girl advertised to be from Venus.

The UFO literature is extensive, and several of the proponents have had financially successful books. Just as von Däniken eclipsed several with priority in the ancient astronaut field, John Wallace Spencer is the star among the UFO writers. A former radio announcer from Massachusetts, he is the author of *Limbo of the Lost*, an attempt to correlate the disappearance of ships and planes in the area of the Atlantic Ocean along the eastern United States seaboard—the so-called Bermuda Triangle—with UFOs. At first he published his book at his own expense, but success brought wider distribution with a commercial publisher. When Spencer lectures throughout the United States, his book sells about 150,000 copies a month. Other books about the Bermuda Triangle, not pushing the UFO theme, are also financially successful. Richard Winer's *The Devil's Triangle*, 1974, sold half a million copies within three months after publication. The account by Charles Berlitz, *The Bermuda Triangle*, 1974, also became a best-seller. A book published in 1975, *The Bermuda Triangle, Mystery Solved*, by Lawrence D. Kusche, analyzing cases of disappearance and reaching logical, natural explanations for each one, was not in the same popular category.

8

Direct Contacts With UFOs

A less-than-critical presentation of the experience of human beings with unidentified flying objects, such as in this and the next chapter, cannot be called the bare facts. The data is more like courtroom testimony without the filter of cross-examination. Much of the material is comparable to hearsay, inadmissable as legal evidence. Thus, John A. Keel, in *Strange Creatures from Time and Space*, describes phantom killers of people and livestock, giant men-monsters with wings, and ghoulish UFO direct contacts as though truth were but a repetition of an experience. The difficult attempt to separate fact from theory, imagination from actuality, and fantasy from reality is absent.

The kind of material is exemplified by the citation in Ian Hobana and Julien Weverbargh's *UFO's From Behind the Iron Curtain*. The authors "report" in their Foreword about an "incident" at the Polish port of Gdynia:

> Several days after the "object" had plunged into the harbour men guarding the beaches met a strange figure which was clearly male dragging himself exhausted along the sand. This creature spoke no known language and was dressed in a "sort of uniform"; a part of his face and hair appeared to

be burnt. The man was taken to the university hospital, isolated and examined. But it was at once apparent that it was impossible to unclothe the creature as the "uniform" had no means of opening. It was not of ordinary material such as wool or leather but of a metal which could only be cut open by means of special tools and after a great deal of effort. The doctors noted that their patient's organs were quite different from ours; the blood system was new to them and the number of fingers and toes was not the normal one. The creature remained alive until a kind of armband was taken off and the "mortal remains were sent for further examination to the Soviet Union."

The author-journalists, instead of checking hospitals and physicians to substantiate or discredit the fantastic tale, chose simply to disseminate the material.

The UFO record is replete with anecdote rather than analyses. Verification and investigation are minimal. Interested and nonaligned people not persuaded by the meager-checked documentation must resort to thinking about the consequences of belief, or those of disbelief, in the reality of UFOs as outlined at the end of the final chapter in this book.

The most outlandish accounts are in *The National Enquirer* type of newspaper and *Offical UFO* type of magazine. In the January 1976 issue of *Offical UFO*, a couple from Indiana described an encounter, 22 October 1975, with two four-foot-high, silver-suited humanoids that "skipped and floated" across a road.

The *National Tattler* special edition on UFOs, Winter 1975, has the story, among others, of the South African woman whose book describes her flying-saucer ride replete with "delicious fruits, vegetables and a pastry similar to our cake." She reported: "They told me the speed of their craft was unlimited and they moved easily and quickly from one solar system to another, but could not travel between galaxies."

Those writers who contend that extraterrestrial visitations have been continuous offer many examples of direct contact between Earth residents and others. W. Raymond Drake cites over two thousand landing reports.[1] In 1967, the *Flying Saucer Review* documented about three hundred. Jacques Vallee lists 923 reliable cases during the century beginning 1868 but also prepared a catalog with 1,247 instances, with 750 of these involving the landing of a craft; more than three hundred had human figures in or about the ship.[2]

Eric Norman gives cases from a variety of historical periods. For early medieval times he cites the Archbishop of Lyons' account of how one ship's crew was captured and stoned to death—this was about 840. During the later Middle Ages, 1207, an event at Bristol, England, serves. A little man dangling from a rope on a spaceship fell to the ground and was attacked by the people. Norman comes to the United States for a late-nineteenth-century example and quotes from the *Houston Post*, 28 April 1897; again, church-goers were involved. They saw a weight being dragged by a rope and looked up to see an airship with several windows and a bright headlight. A small man in a light blue sailor suit descended the rope, but when he saw the onlookers, he cut the rope, and the airship moved out of sight. According to the newspaper, the weight was an anchor "now on exhibition at the blacksmith shop of Elliott and Miller."[3]

Dione, among others, has a similar concoction, and his witness was a farmer and member of the U.S. House of Representatives from Kansas. The report, written in 1897, had many accompanying affadavits and described an airship about three hundred feet long, occupied by strange beings. A cable from the ship hooked on to a two-year-old heifer and hoisted up the animal. Its hide, legs, and head were found in a field the same day.[4]

Bergier's tale about a boy in Wales in 1909 suddenly disappearing on a short walk to a water well offers a small amount of explanation for all historic contacts with UFOs. Bergier speculates, and rejects as untenable, whether a giant condor or a balloon could have taken the boy. Bergier does not mention imaginative tales devised under the influence of alcohol or delusionary narcotics.[5] Clifford Wilson, in *UFOs and Their Mission Impossible,* has other comparable abduction stories, mostly from Otto Binder's *Unsolved Mysteries of the Past.* In 1962, Brazil, a man was taken after refereeing a soccer match; a rubber-plantation worker saw a saucer land, and three humanoids came out to kidnap the referee. In Argentina, during the same year, three sons watched their father being abducted; three days previously the man had come across two little men, about three feet tall, digging a hole. In 1961, a woman parachutist floated down three days after her jump; she said she had been captured by a saucer, was taken aboard, and then released. On 13 August 1965, two women working in a bean field at Renton, Washington, ran from three saucer humanoids with scaly skins and huge protruding eyes.

At the beginning of the twentieth century, Sir Arthur Conan Doyle, immersed in spiritualism, published photographs of tiny, inchlike human figures playing with a young girl. Some English may have been impressed—perhaps the same kind who accepted the stories of American blacks during World War II: that they were night-fighters.

Stories of direct contacts with UFOs became more abundant during the latter twentieth century, after World War II. The ancient astronaut school of writers supply a large number of instances. W. Raymond Drake reports that in August 1947 an Italian painter and writer saw two dwarfs in overalls climb aboard a spaceship in a mountainous region near Venice.[6] Brinsley Le Poer Trench embellishes the tale with the contention that the man had raised his hand for a friendly greeting, and one of the dwarfs had aimed a ray gun to fell him.[7] Drake has another one in Cedric Allingham, living near Forres on the north Scottish

"Extraterrestrial Being" advertisement for cigarette company

coast, who saw and photographed a Martian, 18 February 1954.[8] Eric Norman contributes stories about visitors from Venus. One involved a man who lived near Mt. Palomar, California, the site of the two-hundred-inch reflecting-type telescope. He, George Adamski, claimed contact with a being from Venus. Adamski described the individual as five and one-quarter feet tall, twenty-five to thirty years old, weighing about 135 pounds, with long blond hair and a beardless face. Through mental telepathy, Adamski determined that the Venus people were concerned about radiation from our atomic-bomb testing then taking place. Before his fame as a contactee, Adamski was the head of a mystical cult called the Royal Order of Tibet.

Another in contact with the Venus beings was George King. He was walking in the West End of London in 1954 when he was accosted.[9] R.L. Dione describes how a senior executive of the Steep Rock iron mines, and his wife, in Ontario, 2 July 1950, watched "little men" milling about the deck of a disc-shaped object.[10] Peter Kolosimo adds that an engaged couple in Italy, 2 April 1962, saw a lens-shaped object touch the ground and two figures emerge; one disappeared into a nearby forest, and the other took the vehicle aloft again.[11]

Some of the UFO writers try to be less flamboyant about the direct contacts and even ridicule the encounters. Aimé Michel refers to the "silly yarn told by George Adamski"[12] but, on the very next page, claims that "in November, 1954 in Venezuela . . . there were at least six well-authenticated reports of hairy humanoid dwarfs landing in saucers and encountering human beings."[13]

Drake's stories consistently accent dwarf beings; for example, in 1964 a New York farmer saw some in one-piece suits, and they claimed to be from Mars.[14]

Ivan Sanderson has a similar tale about little visitors. His is about a widow on a Pennsylvania farm who saw trespassers pulling up her vegetables, and they wore grey uniforms and Buck Rogers space helmets. Sanderson also supplies the information that two children's camps were a few miles from the farm.[15]

The auxiliary information needed to assess the incident properly—such as the camps in the Sanderson story—is generally not given; misleading and false data may have even been supplied. Thus, someone picked up a piece of metal, supposedly from a spaceship explosion over Ubatuba, Sao Paulo, Brazil, in 1957, and said it was ultrapure magnesium. Analyses showed the object to be the impure metal. In 1967, the now-defunct *Look* magazine had a special issue on flying saucers, and one photograph showed a "claw-shaped" moist mark on the otherwise dry sand of a beach. Some of the claw-marked sand was analyzed, and urine, probably from a passing bear, was found present.

In the same class of incidents is one reported by Donald E. Keyhoe. He tells of a young girl on a California beach being raped by six humanoid creatures with webbed feet and bluish skin; nine months later the baby also had webbed feet and bluish skin.[16] Drake claims that

> Miss Marlene Travers, age 24, of Melbourne, Australia, alleged that on the night of Aug. 11, 1966, a silvery disk landed from which alighted a tall handsome man wearing a metallic green tunic who told her by telepathy that she had been selected for the honor of being the first woman on Earth to bear a child by a man from his planet. He raped her; a doctor found burns on her arms and legs, later that she was pregnant.[17]

Flindt and Binder claim that in his files, John A. Keel has a large number of sexual-contact cases.[18] The most celebrated case of sexual contact with extraterrestrials is that of a young South American farmer whose escapade was on board a spaceship. His mate was a nude woman with red hair. After intercourse she patted her stomach and pointed to the sky.

Keyhoe has other less extraordinary accounts of direct

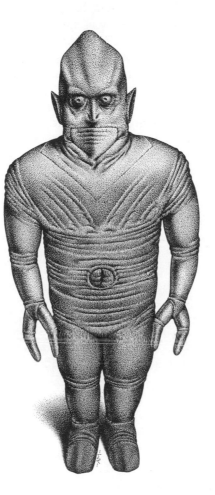

"Space Being" (*Courtesy, WLS-TV, Chicago, Illinois. Artist, Alex Murawski.*)

contacts. He, a former director of NICAP, reports the event at the Brazilian Fort Itaipu, 4 November 1957, when heat from a UFO struck down two soldiers.[19] A similar experience befell James W. Flynn in the Florida Everglades, 14 March 1965. He raised his hand in a friendly gesture, and the parked UFO projected a narrow beam of light knocking him unconscious.[20] Keyhoe's documentation can be questioned if his description of Project Ozma is typical; he has Frank Drake almost immediately detecting intelligent signals from Tau Ceti and the Pentagon clamping secrecy on the experiment.[21]

Many of the contacts are not unfriendly. Around 1950, a technician at the White Sands Proving Ground in New Mexico, Daniel Fry, said he was walking in the desert and came across an "oblate spheroid about thirty feet in diameter at the widest part, and about 16 feet in height, silvery in color." He wanted to touch it, and a voice cautioned, "Better not touch the hull, pal, it's still hot!"[22] Fry discovered that the voice belonged to a spaceman named Alan, who spoke to him via telepathy. Fry was given a saucer ride, described in his book, White Sands Incident. Fry went on to "secure" a Ph.D. degree from a diploma mill and, in 1972, headed, from his home in Merlin, Oregon, an organization called Understanding, Inc. He claimed to be in contact with a member of an "alien race" living in Cairo, Egypt, and operating as an import-export agent.

Peter Kolosimo, on page 76 of Timeless Earth, published in 1973, reports a fantastic tale about a 1959 Soviet mission to Tibet to discover "routes to the stars." An aged lama made two of the Soviets exercise, eat properly, and finally concentrate to see a cloudy image of a man with jointed limbs, standing with a reproduction of a solar system having ten planets. The lama said that one more planet existed beyond Pluto.

Another unbelievable story is that of Woodrow Derenberger of Parkersburg, West Virginia. He met a UFO being named "Indrid Cold," who made numerous visits to Derenberger's home, and sometimes accompanied by a mate.

Another friendly contact was claimed by ex-photographer Gabriel Green, who campaigned for the United States Senate from California in 1962. Green said he knew individuals about four feet tall from the planet Rentan belonging to the star Alpha Centauri.[23]

Similarly, the former science-fiction writer and founder of scientology, L. Ron Hubbard, reports the existence of Thetans. They are from a quarter of an inch to two inches long, arrive in human beings at birth, and are wanderers from planet to planet.[24]

The visitors from elsewhere are friendly enough to do medical miracles. Ralph and Judy Blum have documented several cases.[25] One was a seventy-three-year-old man from Argentina, who grew a third, new set of teeth after noting a powerful light from a UFO. A sheriff's deputy in Texas had an ugly gash on his left index finger instantly healed, and a French doctor lost a limp and ankle pain. The strangest development was the operation by two small UFO occupants on a Brazilian girl dying of stomach cancer. After the half-hour operation, the visitors gave the father—one of seven present—medicine to help the girl's recovery.

The saucer occupants are often friendly enough to reveal their nature to selected people on Earth. Psychologist C.G. Jung, sympathetic to occult phenomena, repeats in his book *Flying Saucers* (p. 119 ff), the message of Orfeo M. Angelucci from his *The Secret of the Saucers*. From one saucer, a man's voice speaking in perfect English said: "Don't be afraid, Orfeo, we are friends!" The voice proclaimed,

We see the individuals of Earth as each one really is, Orfeo, and not as perceived by the limited senses of man. The people of your planet have been under observation for centuries, but have only recently been re-surveyed. Every point of progress in your society is registered with us. We know you as you do not know yourselves.

Clifford Wilson, in *UFOs and Their Mission Impossible* (p. 128), gives the unique story of Albert K. Bender. He had been given a metal disk and by pressing it could contact the spacemen at any time. The disk had disappeared, so Bender felt free to say that the saucers were from another planet; the occupants were after a secret compound in the oceans of the Earth. He was taken to a secret cavern via teleportation, where one being virtually worshipped by others had given Bender the details. Wilson is most impressed with John A. Keel's *UFOs . . . Operation Trojan Horse* and retells from there (p. 130) the story of Reinhold Schmidt, 5 November 1957. He was contacted outside of Kearney, Nebraska, by a saucer occupant speaking in German, who assured Schmidt that he would eventually know about the real meaning of the saucers. Wilson also has the tale first publicized by Donald E. Keyhoe in *Flying Saucers from Outer Space*. Wilson (p. 141–2) describes the Pueblo, Colorado, radio executive who claimed that the United States had seven flying disks, three of them forced down in Montana—and he had been inside one. He gave details: the crew were in cabins pressurized with a mixture of thirty percent oxygen and seventy percent helium; the propulsion was done via electrostatic turbines; different colors observed were due to varying speeds of the saucers.

Free, marvelous travel is also available to believers. Brinsley Le Poer Trench[26] explains several as teleportation—instant transport from one place to another. Thus, on 25 October 1593, a Spanish soldier, whose regiment was in Manila, appeared in Mexico City, and he knew that the governor of the Philippines was dead. Then, an Argentine doctor and his wife found themselves on a dirt road near Mexico City, 4,500 miles from where they had been driving. Group tours have also occurred, with an entire British regiment swallowed up by a "cloud" during the Turkish campaign of World War I. That the ferocity of battle in war brings the slaughter of great numbers is not considered; at least the Turks are alleged to have attempted nearly genocide with the Armenians during World War II.

Cervantes alludes to teleportation in *Don Quixote* (part II, chapter XLI), where the Knight and Sancho, blindfolded, are mounted on Clavilero, the wooden horse, and apparently are whisked through the upper air. But this is acknowledged fiction. The avid believers should also wonder why spacecraft are necessary if teleportation can be so accomplished.

Direct contacts with the UFOs are alleged to be shown by matted-down grass, disturbed soil, and burned vegetation. According to Keyhoe, the Argentine government proclaimed that a UFO landed in La Pamp, 24 May 1962, and two robotlike figures were seen to emerge from the disk. When observed, the figures retreated into the UFO and it rose quickly. The ground was scorched where it had been.[27]

When the ground impressions can be explained away by other factors, in line with the principle of avoiding new explanations when old ones serve well, an imprint by a landing gear cannot be entertained. As long as the ground disturbances can be explained by animal action, electricity, or atmospheric properties, the idea of UFO is an encumbrance.

The idea could be a tentative hypothesis if direct evidence were more substantial. According to Hynek, very few of the accounts of interception by extraterrestrial visitors are in his category of hard evidence.[28] Trained as an astronomer, Hynek accents only those instances where two or more reliable witnesses are available. Actually, two or more people can just as easily be misled as can one; indeed, crowd behavior involves a low-grade type of perception and intelligence.

Hynek recounts the story found in many UFO books of

a priest in charge of the Anglican mission at Boianai, Papua, New Guinea. During the summer of 1959, Father Gill and thirty Papuans saw a large UFO and several smaller ones, and the observers made out four humanlike figures. One waved a return to Father Gill's salute. Donald Menzel explains away the visions as those of the planet Venus seen through a pair of astigmatic and myopic eyes.[29] Philip Klass in chapter 22 of UFOs Explained, published in 1974, notes that the supervisor of Reverend Gill and director of a multistation Anglican mission was an avid UFO buff, reporting and collecting observations for flying-saucer magazines. He enlisted the aid of his missionaries in this venture.

Hynek has many multiple-witness stories. One couple in Leominster, Massachusetts, in 1967, out after midnight to see newly fallen snow, had their automobile lights, radio, and engine stop and the man's pointing arm immobilized by a UFO. Hynek gives a detailed account of a sighting, August 1955, Hopkinsville, Kentucky, having seven adult witnesses; rifle shots against a small "glowing" man with very large eyes were described as "just like I'd shot into a bucket." A third mystery reported occurred in North Dakota, November 1961, where one of four men fired on a human figure near a flying saucer.

Hynek describes the Patrolman Herbert Schirmer, 3 December 1967, Ashland, Nebraska, case even though it is not a multiple-witness one. Author Eric Norman arranged a hypnosis of Schirmer, who then divulged his experience with the UFO people. Schirmer claimed that the beings he had encountered were from another galaxy, although they had bases on our solar system, including the Earth. They were about four and one-half to five feet tall, wore tight-fitting silver grey suits and boots, and had a slit on the face for a mouth.

One good reason for doubting the Schirmer story is that it fits so well into science fiction about space people. The tale has all the embellishments, such as a ship made from

"Just once I wish we could visit a planet that wasn't plagued with UFOs!"
(Courtesy, Grin and Bear It, *by George Lichty—Field Newspaper Syndicate.)*

absolutely pure magnesium, the extraction of electricity from power lines on Earth, a ray gun capable of paralyzing people on Earth, and an electromagnetic force field able to stop automobiles and silence radios. The drama is complete with the UFO occupant informing Schirmer that "your people are very hostile" and a "cover story" being implanted in Schirmer's mind in order for him to report his experience as an ordinary UFO observation.[30]

Those with a great deal of confidence in hypnosis may rebel at the labelling of a confession under hypnotic spell as a fabrication. Yet hypnosis does not reveal the truth—only what has been experienced by an individual, even while dreaming.

The UFO contact by Mr. and Mrs. Barney Hill is highly regarded by believers in the extraterrestrial hypothesis because their adventure with UFO occupants was revealed under hypnosis. Yet it is seldom reported that the white woman married to a black man was a believer in UFOs as extraterrestrial and actually had a dream about being abducted by UFO beings before recounting the story to a psychiatrist.[31] The psychiatrist believed that the Hills had an experience with an unusual aerial phenomenon, but that the abduction story was improbable.[32]

The humanoids described by the Hills were of a different variety. They had black uniforms, holes for ears, and so compressed a mouth and nose that little of it was revealed in a profile view. Their long, slanting eyes gave them a sinister appearance.

During the 1970s, the National Broadcasting Company twice televised a fictional account of the Hill episode. Then, too, an elementary school teacher built a model of the star map allegedly shown to Betty Hill by the humanoids. But several other star alignments, including one centering about the constellation Pegasus, were demonstrated to be similar.

During October 1973, the chief of police of Falkville, Alabama, said he had encountered a spaceman and took a

Polaroid picture. The photograph showed a human-appearing being with no facial features, in a silver suit with what appeared to be an antenna on his head. The chief described the creature as six feet tall and "moving stiff, like a robot" but able to outrun the squad car. The police chief was later ridiculed, divorced by his wife, and fired from his job.

At about the same time a University of Georgia art student told authorities that he saw a sea-green humanlike being emerge from a spacecraft. At Mount Airy, North Carolina, a man said he saw a dwarfed, gold-suited creature.

In October 1973, a forty-two-year-old man and an eighteen-year-old youth were fishing on the Pascagoula River in Mississippi when they saw a strange object emitting a bluish haze. They were taken aboard the craft by three creatures with big eyes, crablike hands, wrinkled skin, and pointed ears, according to their story elicited under hypnosis.[33]

The Mississippi event is one of the chief pieces of evidence presented by believers in extraterrestrial visitation, yet like the Hill episode its veracity depends upon the faith that hypnosis reveals waking and not dream states. The hypnotic condition in itself is a relative mystery, with many unknowns.

The two men involved in the Mississippi adventure were featured in several television shows, and the older one began to prepare a book. However, on the 15 December 1974 NBC television production about UFOs, critic Philip Klass called the affair a gigantic hoax.

During November 1975 a comparable UFO experience was reported by six young men in Arizona. Seven men suddenly saw a vehicle, and one ran towards the object. Apparently, a brilliant blue flash enveloped him. Five days later he returned to claim he had been examined aboard a vehicle by three hairless men in tight-fitting blue clothing. Previous to his "experience" he had asked a local radio station to interview him as an extraterrestrial, and they had refused.

The National Enquirer offered a prize for the best UFO case and by 1 January 1973 had more than one thousand entries. The winner was an event at Delphos, Kansas, on the Johnson farm. Landing marks were difficult to explain away. However, the family operating the farm refused to take lie-detector tests or undergo hypnosis.

UFO organizations accept an incident at Socorro, New Mexico, as the most credible for a direct-contact case. Here a police officer allegedly saw a spaceship land and the outline of creatures. Patrolman Lonnie Zamora first interpreted his sighting as an automobile standing on end, with two children or small adults dressed in white clothing standing nearby. Philip J. Klass studied the situation, including the altered ground, and suspected ball lightning as the real cause.[34]

Several comparable incidents are reported by UFO enthusiasts. Brinsley Le Poer Trench describes a forester in southern Finland, January 1970, observing a UFO, with a nearby creature about three feet tall holding a black box yielding a blinding light. Trench also has the case of a Venezuelan farmer who saw three strange beings near a saucer.[35] The Blums are impressed with the French lavender grower who was immobilized by a pencillike instrument held by one of two with almond-shaped eyes and slits or holes for a mouth.[36] Martin Ebon reports how an elderly lady in England saw two three-foot beings in her raspberry bushes, and they wore helmets as well as snugly fitting suits made of shiny, silvery material.[37]

Repeated stories about dwarfs do not make reality, even though Adolf Hitler's propaganda minister as well as advertising moguls contend that repetition brings the ring of truth to the unsuspecting public. Here, however, there is no plot to deceive; small astronauts have become part of loosely circulated stories. The tales about entertainers who are alcoholic or miserly are in the same category. Those who cannot distinguish between fantasy and reality may also see actuality in the little man of the toilet-cleaner commercial.

The most fantastic contacts are those of the type made famous by George Adamski, extended ones, and now abundant in UFO literature. Adamski was received by Queen Juliana and Prince Bernhard of the Netherlands and given an audience by the Pope; but the others claiming extended communication are known only among the devotees. W. Raymond Drake has the story of Dr. Frank Stranges, head of the International Evangelium Crusades, Inc. of New York, who was told by an inhabitant of the planet Venus how he, the Venus "man," met with the president of the United States and officials of the United Nations. Drake also repeats the tales of Truman Bethurum, who in 1952 met the lady captain of a ship from a planet called Clarion, and Dana Howard who discussed life on Venus many times with "beautiful Diana from Venus."[38]

McWane and Graham sympathetically describe several cases.[39] They cite the Solar Light Center of Central Point, Oregon, where the lady in charge had contact with the Saturn Tribunal, the governing body of the solar system; the Awareness Research Foundation of North Miami, Florida, whose lady director was aboard a Saturn spacecraft; Canadian women also in touch with higher intelligence elsewhere. Brinsley Le Poer Trench has his entry with a West Akron, Ohio, lady escorted into a spaceship by a girl with no facial features and beautiful, dark, chestnut hair.[40]

George Van Tassel calls himself "the Sage of Giant Rock" and holds an annual convention of contactees at his home in California. He started in the field in 1953 when a man introduced himself as "Sol-danda" and invited inspection of

(Courtesy, Industrial Research.)

his spaceship.[41] Photographs of several of those who claim direct contact are in Paris Flammonde's book, *The Age of Flying Saucers.*

If messages from space are considered a direct form of contact, there are a considerable number of individuals on Earth, in and out of asylums, who claim to be recipients. Among those in this category is the Israeli magician and psychic Uri Geller.[42] One described by McWane and Graham, from Cape Charles, Virginia, boasts: "I accomplished 294 miracles! Making three simultaneous hurricanes for a group of scientists; stopping volcanoes in their tracks; controling a giant government radar installation with my mind; controling the city of Cleveland."[43] Robert Emenegger's psychic find is a woman in Maine who tells about a universal association of planets. The delegate from Mercury was in favor of leaving human beings on Earth "to stew in their own juice."[44]

9

UFO Observations

What constitutes a UFO observation varies with the interpreter. Devotees of the phenomenon inclined towards viewing them as having an extraterrestrial origin or beyond the human ken have included as legitimate observations unusual rain of fish, disappearance of planes and ships in the Bermuda Triangle, mutilated cattle in western United States, scorched rings on farmer's fields, manlike birds in West Virginia, big birds in Texas, and investigating and interrogating men in black in black Cadillacs. All these and more are said to be connected to UFOs. Here, however, only sighting of vehicles plus some markings on the ground are considered.

Some of the men and women enthusiastic about interpreting UFOs as extraterrestrial have alleged sightings of long ago. Harold T. Wilkins listed several hundred, beginning 222 B.C.E.[1] George Adamski and Desmond Leslie published a similar tabulation.[2] The quality of these can be estimated by evaluating some mentioned by Norman. He claims that the Emperor Constantine, the Roman ruler who embraced Christianity, saw a cross in the sky in the year 312; that King Richard the Lion-Hearted and King Philip of France saw a "fiery cross in the noonday sky" when discussing their plans to defeat Saladin, the Saracen emperor, England, 1150.[3]

Desmond Leslie traces UFOs to the year 18,617,841 B.C.E., when the first spaceship came from Venus.[4] F.W. Holiday goes only to Neanderthal times, because of markings found on a grave uncovered in 1921.[5] His evidence is more conventional for UFO enthusiasts when pointing to discs in the cave art of Magdalenian culture, twelve thousand to fifteen thousand years ago—such as those at Altamira, Santander Province, Spain.[6]

Jacques Vallee cites a Japanese general who saw moving sky lights, 24 September 1235,[7] while Lore and Deneault assign a UFO observation to Christopher Columbus on the *Santa Maria* enroute to America, 11 October 1492:[8] On 5 April 1800, an object as large as a house passed over Baton Rouge, Louisiana.[9] Aimé Michel reports a "huge flying machine" over Bonham, Texas, 1873, and disks in the Bermudas, Turkey, 1885, New Zealand, 1888, and Oakland, California, 22 November 1896.[10] Andrew Tomas mentions the observation of an explorer in Chinese Turkestan in 1926; a circular-shaped craft was seen through binoculars.[11]

J. Allen Hynek quotes from the March 1904 issue of *Weather Review*, a report from the ship USS *Supply* at sea:

They (lights) appeared beneath the clouds, their color a rather bright red. As they approached the ship they appeared to soar, passing above the broken clouds. After rising above the clouds they appeared to be moving directly away from the earth. The largest had an apparent area of about six suns. It was egg-shaped, the larger end forward. The second was about twice the size of the sun, and the third, about the size of the sun. Their near approach to the surface and the subsequent flight away from the surface appeared to be most remarkable. That they did come below the clouds and soar instead of continuing their southeasterly course is also certain. The lights were in sight over two minutes and were carefully observed by three people whose accounts agree as to details[12]

The problem of what qualifies as an UFO observation has several probable solutions, the least satisfactory being to accept the listings of individual authors, such as those mentioned above. Answers can be obtained from UFO organizations as well as from government-sponsored studies.

The National Investigations Committee on Aerial Phenomena, 1536 Connecticut Ave., N.W., Washington, D.C. 20036, has several thousand observation records on file. In 1964, their then assistant director, Richard Hall, edited and published an 184-page book, *The UFO Evidence*, which lists 746 cases. The U.S. Air Force Project Bluebook files are now in the archives at Maxwell Air Force Base. A complete directory of recent UFOs from these files is reproduced on pages 137 through 189 of *The Official Guide to UFOs*, published by Ace Books in 1968 and edited by the editors of *Science and Mechanics* magazine. Thornton Page, a professor of astronomy, has cited some special observations in the book that he edited with Carl Sagan, *UFOs—A Scientific Debate*, published by Cornell University Press in 1972, as a paperback by W.W. Norton in 1974, and as a record of the discussion on UFOs at the American Association for the Advancement of Science meeting in 1969. The Condon report, the result of the University of Colorado government-sponsored study, has a large number of cases. The analyses was published in hardcover book form by E.P. Dutton and in a paperback edition by Bantam Books in 1969. However, most libraries and book suppliers do not have either, and the report must be obtained from the National Technical Information Service, U.S. Department of Commerce, 5285 Port Royal Road, Springfield, Virginia, 22151. Other sources for UFO information include the Mutual UFO Network, Quincy, Illinois, and Professor Hynek's organization to study the phenomena, headquartered at Evanston, Illinois. In 1975, his center raised $20,000 to support its UFO investigations. Hynek estimates that $100,000 would be needed for a fairly satisfactory job, including computer analyses and field expeditions.

The Project Bluebook files available at Maxwell Field, Alabama, has 12,618 cases. Most of the forty-two cubic feet of material is on microfilm, except for thirty-nine short films, some sound recordings, and a few dozen artifacts.

The large number of sources for recorded observations of UFOs means that questionable sightings can be avoided, unless entertainment is desired. The 18 August 1974 issue of the *National Tattler* reports an Ohio man seeing a disk "manned" by a woman, or at least occupants with long hair.

Enthusiasts for UFOs adversely criticized the Condon report for an alleged avoidance of observations difficult to interpret as meteors, planets, birds, or swamp gas. Yet, insofar as historic sightings are concerned, volume two of the government report, beginning with page 824, documents several, taken from UFO literature, without comment; for example, for August 1566: "People saw a crowd of black balls moving at high speed towards the sun, they made a half turn, collided with one another as if fighting. A large number of them became red and fiery and thereafter they were consumed and the lights went out." And for 6 March 1716: "The astronomer Halley saw an object that illuminated the sky for more than two hours in such a way that he could read a printed text in the light of this object. The time of the observation was 7 P.M. After two hours the brightness of the phenomenon was re-activated 'as if new fuel had been cast in a fire.'"

The kind of observations accented by those inclined to an extraterrestrial interpretation are similar to the ones listed as "unidentified" in the Project Bluebook files. Biologist Frank B. Salisbury, in his article "The Scientist and the UFO," gives some representative sightings of this variety.[13]

The first challenging observation of almost all serious

UFO students was that by Kenneth Arnold, 24 June 1947. He was flying a volunteer search mission for an overdue transport plane and was in the vicinity of Mt. Ranier, Washington. He saw what appeared to be nine disk-shaped objects flying at great speed and in regular formation.

Another interesting sight occurred 16 January 1958. Almiro Baraura, a professional photographer was on board a vessel of the Brazilian navy, *Almirante Saldanha*, in the harbor of Trinidade Island. He saw what seemed to be an object streak behind a mountain peak on the island, make an abrupt turn ,and speed away while emitting a greenish haze.

The so-called Loch Raven case occurred on the evening of 26 October 1958 at a bridge near the Loch Raven Dam, north of Baltimore, Maryland. Two men in a car saw a bright egg-shaped object, about one hundred feet above a bridge. As they approached, their engine stalled, they felt a heat wave, and heard a dull explosion as the object moved up and away.

The incident at Exeter, New Hampshire, 3 September 1965, has been reported fully by writer John G. Fuller.[14] Several witnesses described a huge, noiseless object with brilliant, pulsating lights. Fuller became a convinced advocate, using his column in the *Saturday Review* to express his views.

Equally persuasive for some doubters was the report of astronaut John McDivitt. On the twentieth revolution about the Earth in a four-day flight with astronaut White in 1965, he took motion pictures of a disklike object circling their craft, Gemini 7. Astronaut Frank Borman saw what he referred to as a "bogey," flying in formation with the spacecraft he commanded. The Skylab II astronauts also saw a UFO.

An experienced flyer of combat missions during World War II, sighting a UFO, 8 June 1966, Kansas, Ohio, reported: "There was no sound, and it was as long as a commercial airliner but had no markings. . . . My body reacted as if I had just experienced a 'close shave' with danger. For the remainder of the day I was somewhat emotionally upset."[15] Frightened and fearful observers are not uncommon.[16]

On 16 April 1968, Henry Ford II, the crew members, and some executives on a company plane saw a huge, round object near Austin, Texas. For about an hour, the UFO kept up with the plane, going about 600 miles per hour.

Three governors in the United States claim to have observed UFOs. On 25 April 1966, Governor Hayden Burns of Florida, accompanied by newsmen on his plane, saw one. Several years before he became the chief executive, Georgia governor Jimmy Carter and a group of friends saw a big, shining light in the shape of a flying saucer. The sighting by Ohio's governor, John J. Gilligan, occurred on the evening of 15 October 1973, near Flint, Michigan; his wife also saw an amber-colored beam of light apparently hanging in the sky.

In Africa, President Amin of Uganda, and many others, saw an object in the sky splash into a lake and then fly straight up and out of sight. President Amin declared the event to be a good-luck sign.[17]

Other governments also have similar official and semi-official pronouncements about one or more UFO observations. Peter Kolosimo reports a July 1965 sighting from Argentina:

The naval garrison in Argentinian Antarctic (Deception Island) noticed on July 3rd at 19.14 hours (local time) a huge lens-shaped flying object; it seemed to be solid, of a reddish-green colour chiefly, sometimes changing to a yellow, blue, white or orange shade. The object moved in a zigzag towards the east, but changed course several times toward the west and north at varying speeds and quite silently, passing at 45 degrees over the horizon at a distance of 10 to 15 kilometres from the base. In the course of the movement completed by the object itself it was possible for the eye-witness to get some idea of its enormous speed, and not only because it was poised motionless for about 15 minutes at a height of around 5,000 metres.[18]

Sheila Ostrander and Lynn Schroeder documented some made by Soviet observers, notably science professionals.[19]

Some UFO observations are simply those of strange lights. Project Bluebook analyzed such a case in the experience of Dr. and Mrs. George Walton, traveling north on U.S. 80 in New Mexico. He was a Ph.D. in physical chemistry from Columbia University and had retired to operate a restaurant in Deming, New Mexico. On a clear night with neither moon nor clouds above, they saw at about 10 P.M. what appeared to be a truck heading toward them. They realized that the lights were not exactly on the ground and assumed that it was an airplane on a low-flying mission. The lights came very close and lit up the interior of their automobile. After about a minute or so, the lights fell back and separated; one light went to the northeast and the other to the southeast.[20]

Photographs are available for some situations. Just before noon, 3 August 1965, a highway traffic investigator for the Orange County, California, Road Department, Rex Heflin, took three pictures of an unknown sky object with his Polaroid camera. He tried to contact his home office via radio, but it would not work; once the UFO was gone, the radio operated well.[21]

A majority of UFO observations are at dusk or evening. Lights are inevitably seen together with a disk-shaped object varying in size and color display but almost always noiseless. When the observers are in an automobile, the engine and car lights appear to be adversely affected.

The events as seen from an airplane can be visualized through the experience of Army Captain Lawrence Coyne and three crew members in an army medical helicopter, October 1973. They were on a routine flight in Ohio en route to Cleveland's Hopkins Airport when they spotted a red light directly ahead of them. The light came towards them at a rate of about 700 miles per hour. The helicopter went downwards to avoid collision, but the object also plummeted toward the ground. The UFO positioned itself above the army vehicle and bathed it in a strong, green light. Coyne and his crew reported that they saw a cigar-shaped, grey, metallic UFO, until it slowly accelerated away.

The events from the ground can be visualized by the experience of a family in Columbia, Missouri, 28 June 1973, about 1 A.M. First, two hunting dogs began to whine and retreat; a sixteen-year-old girl saw a light approaching their home. The father was unable to contact the police because the phone was suddenly inoperative. Three people—father, daughter, and baby son—saw a light come closer and retreat a total of three times. When the light, about the size of a house, was near they thought they saw a metallic craft enveloped by a blue glow as well as a pinkish yellow glow. When the object was between two trees in the yard, the trees waved. After about thirty-five minutes, the light receded and disappeared. The dogs stopped whining, and the phone began to operate. Police later found a sizable branch ripped from the trees, and there were some indentations on the ground.

Radar evidence for the existence of UFOs can be cited. James E. McDonald has analyzed several cases involving both visual and radar data.[22] Mass media hardly ever accent these less dramatic instrument sightings. The eight-page special section in the 17 February 1974 National Tattler does not mention radar, while the February 1974 series in Chicago Today has a passing reference to radar.

Analyses of the observations without adopting any explanation has been done by a few investigators. M. Claude Poher, of the rocket division of France's Space Studies Center at Toulouse, did so with one thousand reports. He found that the phenomenon was worldwide and in seventy

percent of the cases witnessed by at least two people. Eight
of ten saw round objects, and the remaining two saw elongated ones. The observers who heard noise were very close
to the UFOs.[23]

"Here—they don't believe me. Say something in Martian."
(Copyright 1975, Universal Press Syndicate.)

The number of UFO observations varies with the year.
In 1947, 122 sightings were officially reported. In 1952—
when jet passenger service began, a hydrogen bomb was
detonated, and Great Britain had its first atomic-bomb
test—1,501 reports were made. In 1957, the year of the first
artificial satellite, Sputnik, more than 1,000 observations were

recorded—twice the amount of the previous four years. In 1966, 1,112 sightings were noted by the Air Force. During August, September, and October of 1973, there were about five hundred observations, more than at any similar length of time in the past. After studying eighteen thousand UFO reports, Dr. David Saunders, formerly a professor of psychology at the University of Colorado, concluded that more sightings were on Wednesday than any other day, while June and September were the most popular months. His studies show that most observations occur during daylight hours.[24] Saunders is also a proponent of the idea suggested by Aimé Michel that a system of grid lines around the Earth delineate the places of UFO reports.[25]

Some enthusiastic for UFOs have written that a larger

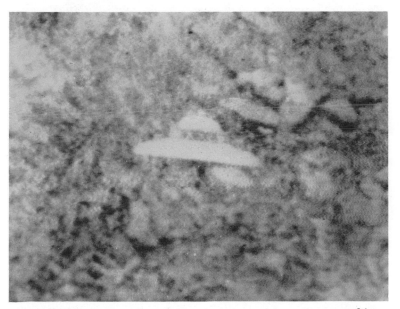

Polaroid photograph of UFO taken by a Lima, Peru, architect, Hugo Luyo Vega. (Courtesy, UPI.)

number of UFO observations occur at times when Mars is closest to the Earth. However, the facts do not support the contention. Another suggested correlation, unreasonable to nonbelievers, is a tiein with events in the Middle East.[26]

Many of the observers are adamant not only about what they saw but also about the interpretation. The dogmatism blocks much of the serious effort to separate perception from theory and to account for the experiences. Clifford Wilson, in *UFOs and Their Mission Impossible*, has a report illustrating the tendency that was first published in *The Sun-Herald* of Sydney, Australia, 23 January 1966. A twenty-seven-year-old banana grower said:

> Had anyone asked me five days ago if I believed in flying saucers I'd have laughed and thought they were nuts. But now I know better. I've actually seen a spaceship. No one will ever convince me that I was imagining things. I was driving the tractor through a neighboring property on my way to my farm about 9 A.M. on Wednesday when I heard a loud hissing noise above the noise engine of the tractor. It sounded like air escaping from a tire. But the tractor tires seemed okay, so I drove on. At first I ignored the sound, but suddenly I saw a spaceship rise at great speed out of a swamp called Horse Shoe Lagoon about 25 yards in front of me. It was blue gray, about 25 feet across and 9 feet high. It spun at a terrific rate as it rose vertically to about 60 feet, and then made a shallow dive and rose sharply. Traveling at a fantastic speed, it headed off in a southwesterly direction. It was out of sight in seconds. I saw no portholes or antennae and there was no sign of life in or about the ship.

Likewise, a Mt. Prospect housewife, writing in *The Chicago Daily News Suburban Week*, 4 September 1975: "I saw a UFO," she claimed, "we are positive of what we saw."

10

Hoaxes and Illusions

The science of paleontology has been a special target for practical jokes and planted fakes. Perhaps the perpetrators had no motive save befuddling the experts. Now astronomy, through UFOs, is becoming a haven for hoaxes. No one can say how many of the UFO observations are in this category, but enough proved cases exist to suspect others where the evidence is lacking; for example, Renato Vesco reports:

> A newspaperman named Perry Heard declared that he encountered four policemen one evening along a deserted road in Missouri. They were taking in a strange dwarf who was babbling in an unknown language into a sort of small oval box. "Perhaps", wrote Heard, "he was a Martian transmitting news of his capture by the Earth people. I wanted to snap a few pictures, but the policemen brusquely ordered me away.[1]

Vesco here quotes a newspaperman who writes conjecture rather than hard news. Investigative reporting, not unfounded supposition, was necessary. During October 1973, newspaper work was more professional, and Jeff and Steve Lowe were found lurking in the woods near Dayton, Ohio. They had frightened some motorists because the boys had

Jeff and Steve Lowe, from Beavercreek Township, Ohio, proved that a 200-foot roll of aluminum foil can make two Martians. (Courtesy, UPI.)

wrapped themselves in aluminum foil and donned masks to appear like beings from elsewhere.[2]

UFO hoaxes began the day modern UFO excitement began in the United States. On 24 June 1947, when Kenneth Arnold saw nine round, silver, metallic objects, flying in formation at supersonic speed, another observer, Harold H. Dahl, while on border patrol in Puget Sound, formulated a tale about five crafts slowly circling around a sixth one that had apparently stopped. He said he saw a barrage of white, metallic fragments fall to the ground. Later, he and his immediate superior admitted that the so-called UFO pieces were "only mineral formations picked up at random . . . they decided to show the press as evidence of the wreckage of the flying saucer with the sole aim of heightening interest in the story."[3]

In 1950 writer Frank Scully had a best-seller with a first printing of forty thousand copies, *Behind the Flying Saucers*.[4] The book had the formulated story of the Pentagon holding bodies of little men whose spacecraft had crashed in the United States. Another perpetrated fantasy was that President Truman several times showed motion pictures of the operation of Venusian interplanetary ships recovered from an Arizona desert.

The first writer to carry through such an elaborate astronomy-related hoax was a reporter for the *New York Sun*. In 1835 he wrote stories about astronomer John Herschel at the Cape of Good Hope finding evidence for life on the moon. He had Herschel claiming the moon inhabitants "averaged four feet in height, were covered, except on the face, with short and glossy copper-clad hair, and had wings composed of a thin membrane, without hair, lying snugly upon their backs from the top of the shoulders to the calves of the legs."[5] The paper tripled its circulation before the truth was uncovered. Communication then was rather slow —via letters on sailing ships—and time elapsed before others queried Herschel and his answer arrived.

During January 1968 the *Denver Post* reported "Thirty Citizens Sight UFO" and printed eyewitness accounts. Then a mother came forward to explain that her two sons, aged fourteen and sixteen, had used a clear plastic, dry-cleaning bag to launch a hot-air balloon, producing the observed effect.

In at least one case the young perpetrator maintained his story for months. A thirteen-year-old boy allegedly took three photos of a UFO and then described the object as a silver disk with a whining noise. Finally, he admitted the fakery, having made a paper-maché flying saucer.[6]

Schoolboys, too, were responsible for the so-called "Ampleforth Abbey" sighting of 1290, reported in the 1953 book of Desmond Leslie and George Adamski, *Flying Saucers Have Landed.* The story was omitted in the 1970 second edition of the book, perhaps because by then the Condon report had shown the tale to be a fabrication by two boys at the Ampleforth Public School.[7]

Youngsters were involved but not responsible for the experience of a scoutmaster near West Palm Beach, Florida, 19 August 1952. He and some scouts saw strange lights in a deserted area, and he went to investigate. According to Dale White in *Is Something Up There?*, the man probably arranged to be found moaning, with hair singed off the rear of a hand, so that he could profit; he had hired a press agent.

Another concoction is that a UFO crashed at Aurora, Texas, 17 April 1897, and the pilot is buried there.[8] In this case, cemetery and historical records are available, and, if need be, actual grave examination can occur.

Astronomer Peter Millman, in his August 1975 article in *The Journal of the Royal Astronomical Society of Canada,* describes the hoax perpetrated by radio officer Z.T. Fogl of the S.S. *Ramsay.* In 1959, the *Flying Saucer Review* published a photo he took, purportedly of a spaceship, and many other periodicals reproduced the item. After Fogl confessed that he had made the picture by first putting together plastic model aircraft parts hung by a silk thread, the *Flying Saucer Review* in 1966 exposed the hoax. Nonetheless, two Canadian journals as late as 1968 showed the photo as genuine.

Hoaxes are generally easier to uncover than illusions. The amount of time necessary to show the true nature of an illusion may be years, and even generations. Cannot the geocentric system wherein the Earth is the center of the solar system be viewed as an illusion? Mankind was engrossed in it until relatively modern days.

Galileo saw the rings of Saturn in such a view that he thought he saw four satellites just as he had observed four Jupiter moons. Every fifteen years Saturn's rings do present themselves in the perspective, tempting the kind of interpretation made by Galileo.

The idea of intelligent life on Mars can be considered an illusion after the Italian astronomer Schiaparelli thought he saw straight-line channels. At the end of the last century, an emeritus professor of astronomy in Charles City College published a book wherein intelligent, civilized life on Mars was called a certainty.[9] In 1924, when Mars was fairly close to the Earth, according to newspaperman Walter Sullivan, "public pressure persuaded both the Chief of Naval Operations and the Director of the U.S. Army Signal Corps to send dispatches to their stations to maintain, insofar as possible, radio silence. . . , in case the Martians tried to communicate to Earth with their more advanced radio technology."[10]

Philosopher Immanuel Kant believed that all the planets were inhabited, with the disposition of the occupants determined by the planet's distance to the sun. Astronomer William Herschel was convinced that intelligent beings lived in a cooler region below the hot surface of the sun.

Astronomers have searched for and have failed to find a planet closer to the sun than Mercury; a satellite of our moon; mountains on the planet Venus; rings for the planet Neptune; and a satellite for the planet Venus.[11] In the light of these events, the UFO experience fits a pattern.

Astronomers, too, albeit not publicly, have thought they saw a UFO. Philip Morrison tells the story of three radio astronomers who saw a noiseless, cigar-shaped object with lighted portholes. However, when the wind shifted they heard airplane engines and recognized that what they had observed had been obscured, and misinterpretation ensued.[12] In the same chapter, Morrison recounts the experience of a renowned historian of aviation who witnessed a motion picture of a UFO and pronounced the object to be some kind of unknown aircraft. Actually, the tail of the plane was photographed through the thick edge of a plastic airplane window, and the record was interpreted as that of a UFO.

Illusions, whether or not viewed as unintentional hoaxes, have been widespread in the annals of UFO observations. Stars, planets, birds, artificial satellites, balloons, and meteorological phenomena have been mistaken for UFOs. In October 1973, Renfroe, Alabama, many citizens reported seeing a UFO and even pointed to a tree where landing occurred. The local sheriff kept the people at a distance while he approached the object. Inside a pouch he found a note reading: "U.S. Department of Commerce, National Weather Service. If found please mail to 451 Ruby Street, Joliet, Illinois."[13] In Longview, Texas, many people saw UFOs land at the airport. The police investigated and found several large geese.[14]

Professor Harley D. Rutledge, a physicist at Southeast Missouri State University, once sent a team of investigators to Poplar Bluff, Missouri, in response to a report of five hundred UFOs flying in formation. The objects proved to be cottonweed seeds caught on a television antenna.[15] Menzel and Boyd in their 1963 book, *The World of Flying Saucers*, describe the lights seen flying in scattered formations over Lubbock, Texas, August 1951, known as the "Lubbock Lights," as the reflection of mercury-vapor lights of the town from the white, oily breasts of plover. Ghostly lights over Norfolk, England, are identified as owls dusted with a common fungus, from their daytime habitat in the trees, which is phosphorescent. Irridescent globes floating in the sunlit sky seen in parts of the United States and France prove to be migrating spiders that spin a balloonlike web, which allows them to be pushed aloft via rising warm air.

Klass describes UFO illusions due to a copper-coated wire reflecting light from the headlamps of a car, airplanes fitted with an intense light beam for night photography, weather balloons seen without binocular magnification, and Venus.[16]

Human sensory imperfections and mind-sets are widespread enough to be responsible for more illusions than people wish to acknowledge. A person may read "Four percent bonds" as "Four perfect blondes," or vice versa; "God hates a quitter" has been read as "Good hats for a quarter." Any out-of-the-ordinary occurrence will inevitably be seen differently by any kind of competent and intelligent group of observers. If a disheveled person were to run into a classroom, pointing a gun and creating a disturbance, two students telling the same story of the affair afterwards would be difficult to find.

Astronomer Peter Millman, in his article in the August 1975 issue of *The Journal of the Royal Astronomical Society of Canada*, reports how visual perception is handicapped immediately by the fact that "there are only some two million nerve channels for carrying the visual images back to the brain. . . . Right at the start, a process of selection occurs. . . . We see what we need to see, what we expect to see, what we are trying to see, what we are conditioned to see."

Perception is also hindered by peer pressure and being taught an abnormal perception by one who believes it to

(Reprinted by permission of Bell-McClure Syndicate.)

be true. Strong personalities, as well as visitors to a group, can resist the accepted interpretation. Psychologist C.G. Jung recounts such an incident in *Flying Saucers*:

> I was once at a spiritualistic seance where four of the five people present saw an object like a moon floating above the abdomen of the medium. They showed me, the fifth person present, exactly where it was, and it was absolutely incomprehensible to them that I could see nothing of the sort. I know of three more cases where certain objects were seen in clear detail (in two of them by two persons, and in the third by one person) and could afterward be proved non-existent.

More recently the magician-entertainer Kreskin, on an Ottawa, Canada, TV show, hynotized fourteen people, and all of them "saw" a UFO at his command.

11

UFO Theory

According to M.G.J. Minnaert, an expert on atmospheric and meteorological optics, about two dozen passengers on an ocean liner in the Indian Ocean at one time saw two suns at low altitude and even took a photograph of the event. No explanation has ever appeared for the oddity.[1] Multiple suns have also been seen in England and described as follows:

> Looking at the sunset sky on May 10 last from about 6.50 P.M. to 7.5 P.M., or perhaps a little later, he saw numberless "suns" in various parts of the sky, coloured "magenta to violet" principally, although they also changed to yellow and green. He called his sister out of the house; after that, his farm manager; and then again referred to the local postman; all of whom verified the curious sight, the lady remarked they (the "suns") looked "like toy air balloons." The locality is near Ivybridge, South Devon. Weather was clear at the time. I understand the "suns" were visible at a moderate altitude, and in the direction of the sunset.[2]

Every natural science has comparable, though not as dramatic, problems, not encompassed by theory. Why then should an explanation for every alleged UFO observation be forthcoming?

The University of Colorado project resulting in the Condon report gave details of several instances where the staff could not propose a presentable idea to cover the event. In their Case 17, South Mountain, Spring, 1967, they stated: "Investigation revealed neither a natural explanation to account for the sighting, nor sufficient evidence to sustain an unconventional hypothesis."

Theories to account for UFO observations are numerous in every category—from natural physical through biological to exotic ones. A leading exponent of natural physical explanations, Donald Menzel of Harvard University gives a long list of possibilities.[3] The idea that the unexplained UFO observations can be accounted for biologically through living forms not known to man has a small number of adherents. Unusual hypotheses, including the supposition of extraterrestrial vehicles, are promulgated largely by nonscientists.

In the Preface to his first book about the topic, Menzel wrote that the sightings not easily explained away were "the rags and tags of meteorological optics: mirages, reflections in mist, refractions and reflection by ice crystals. Some phenomena are probably related to aurora, others are unusual forms of shooting stars. A few . . . probably represent natural phenomena that we still do not fully understand."[4] When testifying before the House Committee on Science and Astronautics, 29 July 1968, he emphasized after-images as a cause. In his most recent accounting, the theme is delineated into several headings, such as material, immaterial, and astronomical objects, but in all events a misinterpretation of the sighting is said to be responsible.

Many attempts to explain away some vexing observations are often patterned after Menzel's thesis. Robert Weedfall, assistant professor of agricultural engineering, state climatologist for West Virginia, and former meteorologist at the Nevada Test Site of the Atomic Energy Commission, seeking to understand the rash of UFO sightings during October 1973, offered the idea of temperature inversion. The month of October had a twelve-day dry period, and air pollutants were trapped in the inversion layers. Distortion and reflection of light occurs with inversions not only in the lower atmosphere but also at the tropopause, the beginning of the stratosphere.[5]

This device is a United States government flying saucer, used in space tests. (Courtesy, Wide World Photos.)

Other scientists have pointed to possibilities not considered by Menzel. Several have attributed much to space junk—mostly parts of artificial satellites—burning up in the Earth's atmosphere. A worker at Mt. Sinai School of Medicine, New York, suggested that small specks of antimatter in the process of annihilation in the Earth's atmosphere was the responsible agent.[6] No one has yet indicated that the daily disappearance of the D layer of the ionosphere (the electrified region of the atmosphere) during the evening hours is connected with the UFO phenomenon.

Soon after 1947, the idea that a potential enemy country, namely USSR, was responsible became popular. No doubt, some Soviet thinkers pointed to the United States in the same manner. During April 1950, radio reporter Henry J. Taylor claimed that flying saucers were highly secret American inventions, and for a short time a satisfactory explanation seemed to be available. In the 1960s, Mel Noel said a group of scientists stationed in Brazil were building UFOs, and he signed up reporters and others for space flights.[7] During the late 1950s, the U.S. Office of Naval Research received an annotated copy of the first of Morris K. Jessup's four books on UFOs. Jessup also received a letter from a Carlos Allende, or Carl Allen, containing allegations that in October 1943, the U.S. Navy conducted a secret experiment, rendering a ship invisible, and teleportation of the vessel occurred; insanity among some of the crew was another result. Perhaps Jessup was Allende or Allen, but the latter was never found, and substantiation of the allegation was not made.[8] However, in 1969, a Carl Allen, or Carlos Allende, confessed perpetrating a hoax, and this was reported in the *Aerial Phenomenon Research Organization Bulletin* (*APRO Bulletin*), July–August 1969.

Philip J. Klass, an aviation writer, contends that ball lightning can account for many observations.[9] (In his second book on the subject, however, he appears to be more inclined to citing a hoax as responsible for the Socorro, New Mexico, incident.) Chemical, electromagnetic, nuclear, and psychological theories have been suggested for ball lightning. The outstanding experimental physicist of the nineteenth century, Michael Faraday, denied the existence of ball lightning, and others have suggested physiological explanations, such as the intense action of ordinary lightning upon the human retina. Perhaps ball lightning, like UFO observations, falls into several categories.

What has been recorded for ball lightning is, however, similar to some UFO experiences. There are vivid, dazzling lights, peculiar smells, an affinity for enclosures, occasional hissing sounds, change of color, and an effect on electrical circuits.[10]

The difficulty with the ball-lightning hypothesis is man's limited experience with ball lightning and the present tiny number of investigators of it. Much more needs to be established. Yet the conception is much more appealing to a scientific frame of mind than is another so-called naturalistic interpretation supported by a few: unknown life forms.

There are at least two claimants for the dubious honor of originating the supposition that some UFOs are a variety of life. According to Renato Vesco,[11] John P. Bessar had the theory in 1946 and told the U.S. Air Force in July 1947. According to Ivan T. Sanderson, the conception is the Wassilko-Serecki theory, after an Austrian titled lady who gave much money and time to finance eccentric biological ideas.[12] Trevor James Constable, author of the book published in 1958, *They Live in the Sky*, calls the material "critters" and recounts in the volume edited by Brad Steiger and John White, *Other Worlds, Other Universes*, an alleged sighting of a life form—presumably a UFO. The story was first published in Ray Palmer's *Flying Saucers* magazine, October 1959. American Don Wood, Jr., flying over the Nevada desert during the 1920s, landed on a mesa where he saw a disabled entity retrieved by a very bright, pulsating, much larger object, and both soared into the sky.

That some UFOs are a form of life is not a solution. Substituting one unknown with another does not promote understanding. The conception may even be divisive and confusing inasmuch as doubt is cast upon the definitions of life, which have borne fruit, Moreover, not a single invest-gator has drawn an inference from the life-form suggestion capable of being observationally or experimentally checked. Sanderson draws an analogy to an organism capable of being reduced to small, dustlike specks and then swelling to normal size, but the citation is far from evidence. Science-fiction writer Gerald Heard, in *Is Another World Watching*, suggested that only insects from Mars would be sufficiently tough enough to withstand sudden turns. His "super-bees," about two inches long, would be "creatures with eyes like brilliant cut diamonds, with a head of sapphire, a thorax of emerald, an abdomen of ruby, wings like opal, legs like topaz—such a body would be worthy of this super mind."[13]

One variation on this theme was concocted by psycholo-gist Wilhelm Reich, a believer in the reality of the UFOs. Both his son, in Peter Reich, *A Book of Dreams*, and his wife, in Ilse Ollendorff Reich, *Wilhelm Reich*, document Reich's convictions. The latter reports (p. 115):

> One of the theories that Reich developed during those years was that the spacemen knew how to use orgone energy, that their machines, their spaceships, were running on orgone energy, and that what Reich called DOR was the offal, the exhaust of their machines.

In the magazine about the bizarre that the late Ivan T. Sanderson edited, *Pursuit*, the journal for the Society for the Investigation of the Unexplained, is one of Sanderson's varia-tions. On page 59 of the July 1971 issue, he proclaims: "Cats, owls, and, as it now appears, many other predaceous animals can see way out into the infra-red and so hunt warm-blooded animals in what appears to us to be total darkness. Some of the smaller wild cats can also see some things we cannot, in bright artificial light." Perhaps UFOs are to be thought of as being in a part of the electromagnetic spectrum not detectable by a human eye and only occa-sionally taking a form so discernible.

John A. Keel has a more macabre variation of the theme

of life forms for UFOs. In *UFO . . . Operation Trojan Horse*
he reports:

> We can speculate that these beings need living energy which
> they can restructure into a physical form. Perhaps that is
> why dogs and animals tend to vanish in flap areas. Perhaps
> the living cells of these animals are somehow used by the
> ultraterrestrials to create forms which we can see and sense
> with our limited perceptions.

Practically all the exotic theories to account for unexplainable UFOs are in the same class of trading one unknown for another. Sanderson postulates something called
ITF, or instant transference, empowering a being to go
anywhere in a flash of time; he thus solves the problem
of interstellar travel.[14] He claims: "We exist in a space-time
continuum that is infinite but not necessarily unique. In
other words, other universes may occupy the same space
at another time, other space at the same time, or the same
space at the same time."[15] R.L. Dione has a so-called fourth
dimension and a space warp; UFOs move instantaneously
from their dimension to ours, and vice versa. In addition,
they are able to adjust their luminosity and color to blend
with the background.[16]

Brinsley Le Poer Trench demands recognition for originating the idea of unseen others living with us in a different
space-time continuum. He complains:

> Although my books were widely read, I was regarded by
> many leading UFO researchers during that period as "off-beat,"
> even to them I was before my time. . . . Today the situation is
> reversed. Leading ufologists . . . recognize that some or all
> of the UFOs *may* emanate from other space-time continuums
> than our own.[17]

However, Air Marshal Sir Victor Goddard, after his involvement in the Royal Air Force investigation of UFOs during
the 1950s, said:

While it may be that some operators of UFO are normally the paraphysical denizens of a planet other than Earth, there is no logical need for this to be so. For if the materiality of UFO is paraphysical (and consequently normally invisible), UFO could more plausibly be creations of an invisible world co-existent with the space of our physical Earth planet than creations in the paraphysical realms of any other planet in the solar system.[18]

Perhaps one can imagine this sort of possibility through the invisible-dog leash sold in magic stores. The device can be held as though an unseen dog were present. Another approach is to toss the item into the parapsychology basket. J. Allen Hynek seemed to indicate such a thought when he told a reporter from *The Trib*, 8 August 1975: "We may be approaching a psychic (paranormal) revelation in the next several decades. It could be that the UFO phenomenon may represent part of that revolution."

The pseudoscience of Eric Norman accounting for UFOs has the added feature of being verbose and meaningless at the same time. He quotes Dr. Meade Layne of the Borderland Science Research Associates, San Diego, California, who used terms such as "emergents," "etheric steel," and "mat and demat."[19]

Brinsley Le Poer Trench speculates about UFOs being organic life forms but nonetheless formulates a conception laden with scientific terms, signifying nothing:

From all this evidence one can now deduce for the first time that flying saucers may be made of silicon and its compounds and the anti-gravity effect is produced and controlled by electric power flowing in the hull and activated by sunlight which is concentrated by a quartz lens on top—the electrical energy may be stored in silicon batteries. The silicon body of the saucer acts as an insulator to gravity and other electromagnetic waves and an alternating current passed through the hull of silicon or through quartz crystals produces a Piezo-

Electric effect or high frequency ultrasonic waves which may be associated with propulsion.[20]

Trench is also a fan of Dr. Mead Layne and uses some of his nomenclature and interpretation. In that spirit, Trench discusses angel hair left by some UFOs as a "gossamer like" material rapidly becoming gelatinous to sublime. "Now it is possible that angel hair and maybe other oily substances could be superfluous materialized energy left over from the materialization of a UFO from another dimension or space-time continuum."[21]

Magnetism appeals to many writers proposing exotic non-explanations. Among them are Gerald Heard and Donald Keyhoe. The latter has more or less repeated the idea in all of his books.[22]

Those infatuated with the occult can also have psychic nonexplanations. Choices run the gamut from the description of the Martian scene by the hysterical medium Helene Smith and her guide Leopold[23] to the vague ramblings of the founder of analytical psychology, Carl G. Jung.[24] In the first case, the girl had simple ideas of Martian language and landscape built out of her imagination. The ideas of Jung have been taken much more seriously. He wrote in *Flying Saucers*:

The plurality of UFOs, then is a projection of a number of psychic images of wholeness which appear in the sky because on the one hand they represent archetypes charged with energy and on the other hand are not recognized as psychic factors. The reason for this is that our present-day consciousness possesses no conceptual categories by means of which it could apprehend the nature of psychic totality. It is still in an archaic state, so to speak, in which apperceptions of this kind do not occur, and accordingly the relevant contents cannot be recognized as psychic factors. Moreover, it is so trained that it must think of such images not as forms inherent in the psyche but as existing somewhere in extra-psychic, metaphysical space, or else as historical facts. When, therefore, the archetype receives from the conditions of the time and from the general psychic situation an additional charge of energy, it cannot, for the reasons I have described, be integrated directly into consciousness, but is forced to manifest itself indirectly in the form of spontaneous projections. The projected image thus appears as an ostensibly physical fact independent of the individual psyche and its nature.

Some see support of such projection in astronomy in the figures that became stellar constellations. Biologist Frank B. Salisbury, in *The Utah UFO Display*, points out a fact that could be used to support Jung's idea, although Salisbury makes no mention of Jung and the theory. According to the biologist, UFO motions, light changes, and other actions are "aimed specifically at the witnesses," tailored to their personalities, social status, and other factors.

The exotic theorists have many speculators in their midst. M.K. Jessup, who had scientific training, believed UFOs were from artificial satellites in permanent orbit at the neutral point of gravity between the Earth and the moon.[25] Gerard O'Neill, a high-energy-particle physicist at Princeton University, worked out a system of placing rotating aluminum cylinders to hold thousands of people at the so-called Lagrangian points—neutral gravitation centers—between Earth and moon. He told a meeting on the colonization of space, held at Princeton University, May 1974:

By 2074 more than 90 percent of the human population could be living in space colonies, with a virtually unlimited clean source of energy for everyday use, an abundance and variety of food and material goods, freedom to travel and independence from large-scale governments. The Earth could become a worldwide park, free of industry, slowly recovering by natural means from the near-death-blow it received from the industrial revolution: A beautiful place to visit for a vacation.[26]

O'Neill estimated the cost of establishing the first colony to be sent to be about 130 billion dollars over a period of fifteen years. In May 1975, one hundred scientists, technicians, and government officials met at Princeton University to discuss the project. During the summer of 1975 another conference was held at the NASA Ames Research Center in California. In 1975, too, scientist Edward S. Gilfillan, Jr. published his scenario *Migration to the Stars*, detailing interstellar travel, an essay spiced with comments on education, psychology, and the human condition.

Other homes cited for UFOs include some of the natural satellites of Jupiter, other planets in our solar system, a planet belonging to another star, and planets in other galaxies. A planet Clarion was invented to meet the need for a home base for the UFOs. It was said to have an orbit between the Earth and the sun and move so that the line between the Earth, sun, and Clarion is always a straight line. Persons on Earth would never see Clarion, permanently eclipsed by the sun. The University of Colorado investigating team assigned this problem to Dr. R.L. Duncombe, director of the Nautical Almanac Office at the U.S. Naval Observatory, Washington, D.C. He studied the problem and concluded there was "definite proof that the presence of such a body could not remain undetected for long."

In 1970, two senior scientists at the Soviet Academy of Sciences proposed what they called a crazy theory: the moon is an abandoned spaceship parked in Earth orbit over two billion years ago. They claimed that their idea accounted for the moon's density being smaller than the Earth's because the moon had a hollow structure. Meteors strike without making a deep crater because the armor-plated hull, covered with loose surface soil, is a protector. Crater floors are convex rather than concave because the armor hull is round. The crazy theory did not gain support —only humorous small talk about finding the door to the inner moon. Nonetheless, in 1976, Dell Publishing introduced Don Wilson's paperback *Our Mysterious Spaceship Moon.*

The purpose of the spaceships is also open to speculation. Are they here to explore the Earth and study, or are they looking for cheap protein in the flora, fauna, and human beings? Could they be members of a peace corps sent to instruct, to keep nations friendly and avoid nuclear holocaust, or are they bent upon conquest of the Earth? Are they looking for a new home because theirs is no longer habitable? Why are they expending great amounts of energy without contacting major sources of information about the Earth?

The appearance of the occupants of the spaceships is not as questionable. Almost all writers have them humanoid, albeit with modifications, such as webbed feet, pointed ears, holes for a nose, or elongated eyes. L. Sprague de Camp and Willy Ley have from a purely logical inquiry arrived at the conception that the most efficient construction for an intelligent life form is something like us.[27]

The exotic theorists argue among themselves. Some scoff at the claim of W. Raymond Drake that "Al Bender an American UFO investigator complained of menaces from three mysterious men in Black allegedly originating from a distant galaxy dominated by an immense burning mass beyond human conception, suggesting a quasar, if life within its proximity be possible."[28]

Drake overwhelms the reader with similar types of "data": Eugenio Siragusa, inspired by beings from Pleiades and Alpha Centauri he met on the slopes of Mt. Etna; Arthur Shuttlewood in Warminster, who received phone calls from aliens from an unknown planet, Aenstria; Bob Rinaud in Massachusetts, who had regular short-wave contact with a planet, Korendor, forty-four light-years away. He credits Dr. G.H. Williamson with the story of a planet, Tyrantor, once a capital of the "Old decadent Empire of the Stars in the Milky Way." The subject planets rebelled, a space war ensued, but the Confederation was victorious; the hopes of "Galactic Peace are still frustrated by evil entities on Orion who have agents here on Earth."[29]

Astronomer Carl Sagan, in a jocular vein, has suggested other possibilities:

Why is there no faction that urges that an unidentified flying object is a projection of mankind's collective unconscious? Psychiatrists have written on the collective unconscious; why not that? Or time travelers? Or visitors from another dimension? Or the halos of angels? Or apparitions from the spirit world, or from the Middle Earth, or Witchland, or Perelandra? There is a wide range of possibilities that could be thought of. How about harbingers of divine wrath? If only we could interpret them properly! Or fulfillments of prophecies from the Bhagavad Gita?[30]

A promoter of the occult has reported a mass hallucination that could be applicable to the UFO phenomenon. Andrija Puharich described the Indian rope trick in terms reminiscent of flying-saucer problems. Several hundred people, including scientists, saw an Indian fakir throw a coil of rope into the air and a small boy climb the rope and disappear. Subsequently, parts of the boy came to the ground. The fakir collected the pieces, went up the rope and both the boy and fakir came down the rope smiling. A motion picture of the event showed only the fakir throwing the rope into the air, it falling to the ground, and the fakir and the boy standing motionless.[31]

John A. Keel in *Jadoo*, reports that P.C. Sorcar, the Indian magician, says that the rope trick is impossible. On the other hand, Keel, when in India, paid to have the trick taught to him.

Mass hallucination may not be the answer for all the UFOs troublesome to explain. The exotic theorists avoid the concept, however, and accent the extraterrestrial hypothesis.

They postulate the mechanics of operation of the vehicles with the introduction of unknowns. Herman Oberth supposes that an artificial field that simulates gravity is generated. The gravitational field affects the surrounding molecules of air pulled along by the vehicle. The latter glides through the atmosphere, protected from the direct action of friction. The occupants, contents, and the craft are in an artificial gravitational field strong enough to overcome any other.[32]

Aimé Michel and others accent the comparable idea of Lieutenant Plantier. While stationed in Algeria during the 1950s, his commanding officer described the appearance and behavior of a UFO to him. Plantier attempted to explain away the so-called four mysteries of the UFOs: their acceleration; resistance to heat; silence; and changes of shape. He assumed the existence of an energy comparable to cosmic rays but not yet detected by man on Earth. The extraterrestrials liberate this energy and create, at the point where it operates, a local field of force that can be varied and directed.[33] He wrote:

It may be imagined that the engine utilizes a method of liberation analogous to that which, in nature, creates the primary cosmic rays. The resulting cosmic corpuscles would radiate through the engine in the direction of propulsion, in the form of a corpuscular-undulatory (particle-wave) fluid moving at a velocity close to that of light. One could thus have a sort of continuous cosmic jet traversing the engine. This jet emitted by the engine would follow it in its movements, propelling it, and supporting it when it was stationary, somewhat in the fashion of a ping-pong ball supported by a jet of water.[34]

Proponents of the three types of theories—natural and within our purview, unknown life forms, and exotic extraterrestrial ones—seldom debate their views. Professor Menzel ignores the last two kinds of concepts, while the extraterrestrial advocates occasionally shoot their arrows of contumely.

Bergier mocks the orthodox astronomy point-of-view of the objects entering our atmosphere on the night of 9 February 1913. He cites Professor C.A. Chant of the University of Toronto observing luminous bodies traveling in a group—first three and then two; they were low enough to cause sonic booms. W.F. Denning reported to the Royal Astronomical Society of Canada that it was like an express train with lighted windows in the sky. Bergier does not believe that the objects were meteors because once one enters the Earth's atmosphere it inevitably falls or is sublimated, and the objects did not.[35] However, on 10 August 1972, many observers in western United States and Canada saw an extraordinarily large and brilliant meteor that reached a point of only thirty-six miles above eastern Idaho. The meteor narrowly missed falling to the Earth, and it zoomed out of the Earth's orbit.

Aimé Michel attacks the ideas of Professor Menzel. Michel is particularly contemptuous in the case of a pilot who fought with a luminous and apparently nonmaterial globe on the evening of 1 October 1948; Menzel reported: "I think that Gorman was right when he stated that the foo fighter seemed to be controlled by thought. However, the thought that controlled it was his own. But the object was only light reflected from a distant source by a whirlpool of air over one wing of the plane."[36] Michel points to several witnesses who saw the light go away at tremendous speed at the end of the aerial duel.

The believers in UFOs as life forms or as extraterrestrial vehicles would like their theories to be more adequately tested; Keyhoe, for example, proposes a lure be set for the spaceships so that they would land and he observed in an unmolested manner. He would set aside an airport with hidden investigators for this purpose.[37] Perhaps a substitute could be the band of renegade Mormons in Mexico who sit naked on their housetops waiting for the spaceships.[38]

The basic problem, however, is not settled by the test of a theory. Many of the proponents of the UFO phenomenon appear to want an explanation of every experience, whether or not others term it dream, misperception, or vision, and much data is missing. Author Daniel Cohen reports in *The Ancient Visitors* his investigation of a sighting where allegedly many people were witnesses. He found so-called observers who denied the event and referred to the UFO publicist as a "nut." Yet the particular UFO story involved keeps circulating.

12

The Basis For Belief

If science can be taken as the arbiter of the UFO puzzle, a distinction must be made between scientific and engineering people. The April 1971 issue of *Industrial Research* showed that eighty percent of those who responded to the magazine's poll rejected the Condon report, and thirty-two percent believed UFOs came from other worlds. On the other hand, Major Donald E. Keyhoe lists in the appendix of his book, published in 1973, only twenty-six scientists and engineers calling for a new, unbiased investigation of UFOs.[1] In 1977, Peter Sturrock of Stanford University, a strong supporter of a renewed investigation of UFOs, polled the 2,611 members of the American Astronomical Society. About half, 1,356, replied, and about 80 percent checked "certainly," "probably," or "possibly" for renewed study of UFOs. Sixty-two respondents even claimed to have seen a UFO. Thirteen members had decidedly negative opinions. One wrote: "I object to being quizzed about this obvious nonsense."

On 21 October 1966, *Science* printed a letter from J. Allen Hynek, the astronomer long in charge of Project Bluebook as scientific consultant to the Air Force. He wrote:

Despite the fact that the great majority of reports resulted from misidentifications of otherwise familiar things, my own concern and sense of personal responsibility have increased and caused me to urge the initiation of a meaningful scientific investigation of the residue of puzzling UFO cases by physical and social scientists.

In his letter he also noted some other conclusions: truly puzzling reports came from stable, educated nonzealots about the issue, including those scientifically trained.

Hynek has never publicly supported the extraterrestrial hypothesis, but several scientists have. Professor Charles A. Maney, emeritus professor of physics and mathematics, Defiance College, Defiance, Ohio, as well as the internationally known astronautics specialist Professor Herman Oberth, long a believer in the occult, are reported to be adherents by Loftin.[2]

Other scientists who have been or are public supporters of the extraterrestrial hypothesis include Frank Salisbury, head of the Plant Science Department at Utah State University; Morris K. Jessup, an astronomer; and nuclear scientist Stanton T. Friedman. He suggests that the name UFO be changed to EEM for Earth Excursion Module.[3] In 1967, his fee for a UFO lecture was $300 plus round-trip air fare from Pittsburgh; in 1973, for allegedly more money, about two thousand came to the thirteen-hundred-seat auditorium of the Mississippi State University at Starkville to hear him.

Some adherents among the scientist supporters would object to his inclusion; but Timothy Leary, dismissed as a lecturer in psychology at Harvard University in 1961 and active in drug-culture circles, is also a believer. He says he has been receiving messages from intelligent life elsewhere in the universe.

The single most important scientist advancing the idea that some UFOs come from other worlds was the late Professor James E. McDonald, senior physicist in the Institute of Atmospheric Physics and professor of meteorology and climatology at the University of Arizona. (Both Jessup and he were suicides.) McDonald stated that the explanations of UFOs fall into eight categories: hoaxes, fabrications, and frauds; hallucinations, mass hysteria, and rumor phenomena; misinterpretations of well-known meteorological, astronomical, and optical phenomena; advanced technologies on Earth, such as satellite reentry; poorly understood physical phenomena; poorly understood psychological phenomena; extraterrestrial probes; and messengers of salvation and occult truth. He agreed that the first four categories account for a substantial number of observations. He ignored the last item listed; felt that numbers five and six were difficult to handle; and so came to the extraterrestrial hypothesis as the least unsatisfactory one to account for a sizeable residue of unexplained observations from credible witnesses.

When McDonald addressed the United Nations Outer Space Affairs Group, he said that even if UFOs were found not to be of extraterrestrial origin, "the alternative hypotheses that will demand consideration will even be more bizarre, and perhaps of even greater scientific interest to all mankind."

McDonald was in the forefront of scientists pressuring the government for an investigation of UFOs. Others clamoring for a substantial study, such as NICAP members, were aided by the series of sightings in 1966, culminating in a famous swamp gas explanation for an experience in Michigan.[4] In October 1966, the Air Force made a $313,000 grant to the University of Colorado to make an independent study of UFOs. The university reluctantly agreed to the assignment, and the director of the study, physicist Edward U. Condon, told the Air Force when he accepted the post that the result would "not necessarily contribute to the nation's peace of mind."

The investigation was attacked before it really began. Dr. McDonald stated that "the Colorado program is not nearly large enough to cope with the apparent dimensions of this problem." He said, "even if the Colorado program would quadruple its scientific staff . . . I would still be saying that we must get more good people onto this problem. . . . Only a large increase in high-caliber scientific manpower attacking the UFO enigma will suffice to make real progress on it."

At first, NICAP cooperated with the University of Colorado study. Then they became annoyed at Condon's public posture and did not release their files. According to Keyhoe,[5] Condon was generally gestureless at meetings but chuckled at the story of the farmer in Brazil who claimed to have had sexual intercourse with a space woman from a UFO.[6] Condon did tell a *Look* interviewer preparing their 1967 UFO special issue: "I won't believe in outer space saucers until I see one, touch one, get inside one, haul it into a laboratory and get some competent people to go over it with me. . . . I would love to capture one. After all, that would be the discovery of the century—the discovery of many centuries—of the millenia, I suppose."

The Condon study had its own organizational problems. The director had to reprimand some, and ex-employees rushed to print a challenge.[7]

The McDonald-type critics evidently wanted a report similar in nature to the Vallee books.[8] Vallee is an astronomer and mathematician, at one time an associate of J. Allen Hynek. The Condon group more or less made note of the sentiment in one of their conclusions:

> Our conclusion that study of UFO reports is not likely to advance science will not be uncritically accepted by them (scientists). Nor should it be, nor do we wish it to be. For scientists, it is our hope that the detailed analytical presentation of what we were able to do, and what we were unable to do, will assist them in deciding whether or not they agree with our conclusions. Our hope is that the details of this report will help other scientists in seeing what the problems are and the difficulties of coping with them.

The UFO enthusiasts did have their day—29 July 1968—in a congressional "Symposium on Unidentified Flying Objects," organized principally by Congressman J. Edward Roush of Indiana. The experts assembled were so biased towards one view that Professor Carl Sagan remarked, "I might mention that, on this symposium, there are no individuals who strongly disbelieve in the extraterrestrial origin of UFOs and therefore there is a certain view—not necessarily one I strongly agree with—but there is a certain view this committee is not hearing today." A summary of the hearings was published by John G. Fuller in *Aliens in the Skies.*

The UFO problem did not diminish in stature with the publication of the Condon report. Instead, old fires were rekindled, and the American Association for the Advancement of Science became involved. A panel consisting of physicist Philip Morrison, astronomer Thornton Page, physicist Walter Orr Roberts and astronomer Carl Sagan arranged for a discussion of UFOs at the December 1969 meeting. It was no easy task. Conservative astronomers persuaded the section on astronomy to resolve against its being held, and some members approached the vice-president of the United States to attempt to bar the meeting.

At the Boston meeting, James E. McDonald and J. Allen Hynek were supported in part by film analyst Robert M.L. Baker, Jr. and sociologist Robert L. Hall. The other side was led by Donald H. Menzel, and he was helped by astronomer Frank D. Drake, psychiatrists Lester Grinspoon and Allen D. Persky, astronomer William K. Hartmann, Philip Morrison, and Carl Sagan. Condon did not speak but contributed an essay to the printed version of the meeting.[9]

The AAAS meeting intended to clarify rather than polarize, persuade instead of provoke. Nonetheless, scientists who support the extraterrestrial hypothesis still cite the mistake of the French Academy of Sciences who dismissed stories about stones that fell from the sky. The great French chemist Antoine Lavoisier reported that a meteorite he had examined was terrestrial, and enlightened Thomas Jefferson said, "I would rather believe that those . . . Yankee professors would lie than believe that stones would fall from heaven."[10] Professor Hynek pleads that case, also popular with the ancient astronaut people: "The history of science is replete with "explaining away" in order to preserve the *status quo*: Discovery of fossils of extinct species, pointing strongly to the concept of biological evolution, was met with many contrived attempts to demolish the fossil-fingers pointing unmistakably to Darwinian evolution. Many, too, were the pat explanations, before facts finally demanded the acceptance of the theory of circulation of blood, the heliocentric hypothesis, hypnotism, meteorites, disease-causing bacteria, and many other phenomena that are accepted today."[11]

The innocent bystander impressed with the credentials of the scientist believers in the extraterrestrial hypothesis need indeed go to the annals of the history of science for proper balance and perspective. Every generation has had distinguished scientists eagerly supporting outlandish ideas; for example, at the turn of the century, outstanding biologist Alfred Russel Wallace was a proponent of spiritualism and phrenology; the celebrated physicist William Crookes fervently believed in spiritualism; in our day, top-notch physiologist Andrew Ivy gave up his career in support of a worthless liquid, krebiozen, purported to cure cancer.

A more significant criticism of Hynek in particular is his avowed espousal of the scientific habit of thought but his failure in one important instance to practice it. In his UFO book he writes:

I have attended many gatherings of scientists, both formal and informal, at which the subject of UFOs has been brought up incidentally, either by chance or sometimes "innocently" by me in order to observe the reaction. I have found it amusing thus to set a cat among pigeons, for the reaction has been keeping with the traditional "weigh and consider" stance of mature scientists.[12]

Yet when asked his opinion of the Philip J. Klass thesis that ball lightning could explain the elusive UFO phenomenon, Hynek responded, "Klass dismissed." In the June 1975 *Ladies Home Journal* he is also quoted as saying, about the investigation of UFOs: "We've had a quarter of a century of buffoonery. Now its time to cut the comedy and get to work." Hynek emphasizes in his book: "No scientist would willfully allow ridicule to be an accepted part of his scientific method."[13] However, Hynek wrote an article for the *Science Teacher,* printed in the December 1974 issue, urging an open mind about UFOs.

UFO study has become a part of higher education. In 1973, the University of Wisconsin awarded a Ph.D. degree for the thesis, "The Controversy Over Unidentified Flying Objects in America: 1896–1973," published by Indiana University Press in 1975, as David Michael Jacobs' *The UFO Controversy in America.* In 1974 Roosevelt University began to offer a course on the subject. It is the academic people, too, who have investigated and analyzed the contentions of the ufologists.

One scientist has asserted that the alleged spacecraft could not be under extraterrestrial control because the laws of physics would be violated.[14] Physicist Hong-Yee Chiu has calculated how much material would be required for only one UFO per year to the Earth since the time our galaxy began and found that half a million stars would have had to be processed for the needed metals.[15] Astronomer Carl Sagan, replying to an inquiry from *Science News,* said: "The remarkable thing about UFO stories is that there are many interesting reports which are unreliable; there are many reliable reports which are not very interesting; but I don't know of any that are both interesting and reliable."[16]

Definitive conclusions are accepted in the scientific enterprise when corroborated evidence is available. Those who reject the extraterrestrial hypothesis complain about the lack of substantial material to warrant the assumption. They cannot go from a belief that life is abundant in the universe to the idea that some life is highly advanced and visiting us. Much in-between factual data, not now available, is necessary for the transition.

The unexplained UFO observations are not compelling because every area of knowledge in every discipline has a percentage of data not encompassed by the prevailing theories. Only the second law of thermodynamics seems to cover perfectly all within its purview; it is an excellent law of description, in contrast to the enormous number of laws of explanation in science. If every branch of natural science has its fund of inexplicable, why should the UFO realm be an exception and have instant explanation of the so-called difficult cases ?

The scientific workers who are not enthusiastic about the McDonald-type views do not reject the conception because it is unconventional. At one physicist's gathering, a celebrated scientist, commenting about another matter, said that the theory proposed was not "crazy enough." Scientists must accept or reject an idea on the basis of its efficiency, not its appearance, history, origin, or promulgators.

Long ago, thinkers used the criteria of elegance, beauty, and simplicity with which to judge ideas. Were they in the realm of religion, they also applied the criterion of compatibility with established concepts. Today, too, scientists also have aesthetic and efficiency values in their work. They do not introduce another idea when an existing one serves well; they do not overthrow useful concepts at a whim.

Physicist Max Planck, in his scientific autobiography, said that ideas in science change not because better ones appear but because the oldsters die and young people take over. A generation of UFO controversy has passed, and the new generation of scientists appear to be no more willing to adopt the extraterrestrial hypothesis.

The ufologists may have a variety of reasons for adopting the extraterrestrial concept. Perhaps one may be paranoiac and have alarmist tendencies, such as the kind shown after the Orson Welles broadcast in 1938 about a landing of people from Mars. Some sailors noting the firelike appearance of sails, now called St. Elmo's fire, and due to electrostatic phenomena, displayed the tendency: they even jumped overboard.

In England the paranoia has taken a practical turn. Members of the British Civil and Public Services Association asked the government to prepare for the arrival of extraterrestrials. The union wanted a "firm statement of policy from the government on its attitude towards such visitors, with special regard to employment and social security benefits." In the United States, the "spaceman" of San Diego, California, diagnosed as a paranoid schizophrenic, wears a suit that is made in part of aluminum foil. He says that he is from a distant planet and has been sent to save the earth from destruction.

Robert Loftin writes:

Why do we desperately desire to escape the fact that we are not alone in this universe? Is it that we are so preoccupied with problems that seem more close at hand, more important to our survival? Has the present civil unrest, crime and corruption, coupled with the war in Vietnam deadened our senses to still further dangers ahead? The leaders of our great society would be wise to take their heads out of the sand and face up to the UFO threat, or suffer the consequences.[17]

Mort Young also reveals the trait:

When I first became entangled in this subject I began experiencing the first of a number of phases which, I have since learned, everyone who digs deeply enough into the UFO mystery passes through. . . . First, there is sheer wonderment. Then comes the fear. It is not quite fear of the unknown or the unexpected. It is fear of the suspected. . . . This leads, with almost unnatural ease, into the third phase; that of suspicion. The victim suspects everyone. He suspects officials, he suspects persons he has met in the search for the final answer. And then he suspects himself.[18]

Major Keyhoe, disgusted with Air Force bureaucracy, falls into the pattern. He finds a connection between UFOs and airline crashes, the great Northeast United States blackout of 1965, and purports damage can accrue to Earth people from UFO radioactivity.[19] He even sees UFOs sparking World War III. John A. Keel, in *UFOs . . . Operation Trojan Horse*, takes a similar line, writing: "Suppose a plan is to process millions of people and then at some future date trigger all of those minds at one time? Would we suddenly have a world of saints? Or would we have a world of armed maniacs shooting at one another from bell towers?" In *The Mothman Prophecies* he continues the Fortean tale of human beings outwitted by the extraterrestrials.

The writers, such as Keel, who promote the men-in-black concept, the idea that extraterrestrials have agents on Earth dressed in black and who travel around in black Cadillacs, are also exploiting the paranoia. The men in black are supposed to be destroying evidence and frightening observers into silence. According to the *National Enquirer*, late September 1975, a sociology professor at Eastern Michigan University claims the government knows about flying saucers being extraterrestrial, and the men in black are from the Central Intelligence Agency.

In a community beset with alienation, distrust, fabrication, and cynicism, it is easy to assume the attitude that the government, Air Force, or both are hiding information or purposely misleading people. In many sections of the United States, as well as in certain underdeveloped countries, people can be found who believe that man's moon-landings, beginning in 1969, are a hoax. The drama of Watergate is enough to mold skepticism about any public organization. Did not the Department of Defense deny and then confess weather-modification experiments in Viet Nam? Perhaps the argument would be that officials are withholding the truth about UFOs.

A good way to avoid the entrapment of alarmist paranoia

is to make a cost-and-benefit analysis for more investigation of the phenomena. A generation of reports and more than half a million dollars in public money has not shown a probable harm to Earth residents by not having a good explanation for the hard-core UFOs.

Perhaps many embrace the extraterrestrial belief for personal-security reasons. With the concept, they are no longer responsible for their behaviour because the so-called advanced intelligences operating the saucers also control people of Earth. They have found a new great father that they can appeal to and worship. The gods of the flying saucers give medical aid, if some fantastic stories are to be believed, and the disks represent the savior who releases Earth beings from individual responsibilities. In Newport, Oregon, October 1975, twenty people were induced to give away all their possessions, including children, and prepare to leave the Earth in a UFO. By January 1976, many of these, now disillusioned, were preparing to return to society, at a "halfway house" established for them in Los Angeles by a lady who had followed the alleged two leaders—a man and a woman —in May and June 1975.

Earlier societies had their kinds of security blankets. The Incas would convince themselves that "I am an Inca; I am not afraid." More recently, Hitler and cohorts in Germany tried a similar kind of indoctrination.

Traditional religion need not be forsaken in the ufologist movement. Already the Christian scriptures have been reinterpreted.

Differentiation between fantasy and reality need not be a concern: men and women can be fervent believers of anything; for example, some people able to cope with daily stress and problems are certain of the existence of inch-high humanoids, such as the characters in Mary Norton's creation, *The Borrowers*. How else can vampires, the dead who suck the blood of living beings, be explained?[20]

Jacques and Janine Vallee, on page 286 of the 1974 paper-

back edition of *Challenge to Science: The UFO Enigma,*
cite the address before the American Sociological Association of Professor H. Taylor Buckner of the University of
California, Berkeley. He proposed that people who believed
in the flying-saucer direct-contact stories were already func-
tioning in the world of the occult. Eighty percent of the
"Saucer Club" audiences, he said, were older women without much formal education, and a belief that "seeing things"
was their unique province.

Perhaps some believers are driven to the idea of other
worlds in contact with us because of dissatisfaction or some
kind of failure. Colin Wilson in *The Occult,* p. 109–110,
claims that "all human beings share a common craving: to
escape the narrowness of their lives, the suffocation of their
immediate surroundings . . . 'magic' or occultism is a simple,
direct method of escaping the narrowness of everydayness".

A sociological view was described by Donald I. Warren
in his article "Status Inconsistency Theory and Flying Saucer Sightings," *Science,* 6 November 1970. While not denying
possible reality to UFOs, Warren reported that status inconsistent individuals, those whose income, education, and
occupation are not at levels consistent with each other, are
prone to UFO experiences. He found that white, status
inconsistent males of moderate income were sixteen times
as likely to report a UFO sighting as comparable white
consistent males. He wrote: "The UFOs provide, therefore,
a form of escape into unrealized and perhaps unrealizable
consistent situations."

Martyrs appear to reinforce concepts, and ufology has
its share in those who suffer ridicule and personal disaster
because of the belief in or experience with a UFO. More
than one peron who publiclv claimed some kind of contact
has been divorced, lost a job, and suffered ill health. Their
only hope of recouping part of their loss is to collect a prize
from the *National Enquirer* for the best UFO experience,
or the $10,000 Philip Klass offers, or the $1,000,000 the

National Enquirer offers for valid proof of the existence of an extraterrestrial UFO.

The cultural climate, too, aids and abets the movement. In times when reason appears to flounder and fail, the occult rises, albeit the UFO group would object to the classification of occult or fantasy. Even if the category is questioned, the seeming inability of reason to have quick answers for pollution, war, excesses of technology, overpopulation, crowding, crimes, corruption, racism, poverty, and the other problems of Western urban civilization drives many to some· kind of belief: UFOs are readily available.

Book publishers and TV producers should share some of the responsibility. Large numbers of ancient astronaut and UFO volumes are available, but there are very few books explaining away the extraterrestrial theme; there are very few exposés of the distortions, misrepresentations, fabrications, and hoaxes. The public has available much pseudoscience, but not much science. Mass media, too, are responsible for accenting the bizarre in their quest for more readers, listeners, and viewers.

The ancient astronaut idea, as well as the extraterrestrial hypothesis for UFOs, is simple, needing no arduous study for understanding. With them, the nondiscerning public can be in touch with the forces of nature. The abstractions of quantum and relativity theories need no longer veritably shut the average person out of physical science.

Some men, women, and children seem to reject the scientific refusal to build beliefs based upon an unknown and not to be perturbed by a segment of unexplainable phenomena harming nobody. Evidently, these people yearn for some kind of intervention from beyond and cast aside evidence. As one of the consultants to the Condon study remarked about his conclusion that a planet called Clarion was not in our solar system: "I am afraid it will not change the minds of those who believe." A group from the Laboratory for Research in Social Relations at the University of

Minnesota in their Festinger, Riecken, and Schacter *When Prophecy Fails* reported a similar conclusion about a comparable UFO situation studied in some depth. Failure to change minds may come, too, from the antiscience bias of some believers. John A. Keel wrote in *Strange Creatures from Time and Space*:

> There exists a large and vocal group of men who *are* unreliable and often irresponsible. Over the past several years our work has brought us into almost constant contact with this group. They call themselves "scientists" and they usually put a Ph.D. after their names. Science has become a sacred cow in this generation but that term is a misnomer. The gender is wrong. Science, by and large, is a lot of bull.

Psychologist Albert Ellis notes other reasons why some need to believe in the mystical: the need for certainty, ego-raising desires, and unwillingness to accept human restrictions.[21] Some ufologists may have such drives. Some may be captured by the fantasy in the same way that alcohol or drugs have ensnared other people. Also, many in our society, imbued with television, movies, and newspaper images, fail to make a distinction between image and reality. Image-making is a billion-dollar enterprise in Western society, and men, women, and children may easily be attuned to images.

UFOs offer an escape and a new life. In 1959, an Austrian court sentenced a swindler to jail for extracting money from people he had "chosen" to settle on Venus.

What is the average nonscientist to do when bombarded with the extraterrestrial or another exotic hypothesis? He or she may not have the scientific frame of mind to make a proper assessment. Despite the cases where scientists have been mistaken, established science can be taken as the authority in this situation.

Tell-tale signs can be used to discriminate when pseudoscience masquerades as science. If, for example, a writer uses the phrase "advance claim" instead of "prediction," he becomes suspect. Sanderson, a biologist, invites the criticism because he describes to invoke mystery rather than clarification. In one of his stories he tells about a being, seven feet tall, posing as an insurance examiner, flashing a gold badge on his shirt, and driving in a "large black car, without lights."[22] Loftin is in the category because he reports a patent untruth about a wrecked flying saucer under control of the Norwegian General Staff, and a flying saucer knocked to the Earth by a hydrogen bomb.[23]

Pseudoscience enters the college classroom, too. A guest lecturer at the University of Denver, 8 March 1950, told about a crashed flying saucer and the little men who had burned. At one landing, the little men were rapid runners and evaded capture.[24]

Pseudoscience is typified, too, by a liberal sprinkling of scientific terms and even by some latest scientific achievement in support of a dubious idea. There is no check of veracity, and concepts already discarded are presented as correct. W. Raymond Drake is a pseudoscientist when he attributes signals received from CTA–102 as being emitted by extraterrestrials, and pulsars as artificial.[25]

Generations ago, authorities in church and state told most people what to believe. Heretics found their truth through what they called divine inspiration, study, or competing points-of-view in other communities. The situation today differs only in the nature of the authority and the greater number of avenues open to the nonbeliever. In every Western society, men, women, and children believe what they do because of parental influence, school indoctrination, peer pressure, religious conviction, salesmanship by individuals and groups, faith, prejudice, and self-interest. Until very recently, scientific conclusion was high on the list of authority for modern people. Its reemphasis as the best guide to belief and action is the solution to the UFO puzzle.

Notes

Chapter 1

1. Erich von Däniken, *Chariots of the Gods?* (New York: Bantam Books, 1971), p. 27.
2. Erich von Däniken, *Gods from Outer Space* (New York: Bantam Books, 1972), p. 67.
3. Eric Norman, *Gods, Demons and Space Chariots* (New York: Lancer, 1970), pp. 45–6.
4. Alan and Sally Landsburg, *In Search of Ancient Mysteries* (New York: Bantam Books, 1974), p. 158.
5. Andrew Tomas, *We Are Not the First* (New York: Bantam Books, 1973), p. 1.
6. Jean Sendy, *Those Gods Who Made Heaven and Earth* (New York: Berkley, 1970), p. 30.
7. Jean Sendy, *The Coming of the Gods* (New York: Berkley, 1973), p. 34.
8. Francois Bordes, "Mousterian Cultures in France," *Science*, 134, 22 September 1961, p. 810.
9 Friedrich Klemm, *A History of Western Technology* (London: George Allen and Unwin, 1959), p. 17.
10. Andrew Tomas, *The Home of the Gods* (New York: Berkley, 1974), p. 60.
11. June Goodfield, "The Tunnel of Eupalinus," *Scientific American*, 210, June 1964, pp. 104–12.
12. Charles Berlitz, *Mysteries From Forgotten Worlds* (New York: Doubleday, 1972), p. 42.
13. Robert Wernick, *The Monument Builders* (New York: Time-Life Books, 1973), p. 5.
14. C. Singer, A. Holmyard, and R. Hall, eds., *A History of Western*

Technology (New York: Oxford University Press, v. 1, 1955), p. 8.
15. *Science,* 8 June 1973, 180, p. 1076.
16. Thor Heyerdahl, *Aku-Aku, The Secret of Easter Island* (London: George Allen and Unwin, 1958), p. 121.
17. Robert F. Heizer, "Ancient Heavy Transport, Methods and Achievements," *Science,* 153, 19 August 1966, p. 821.
18. Lyall Watson, *Supernature* (New York: Doubleday, 1973), pp. 177–8.
19. Tomas, *The Home of the Gods,* pp. 89–90.
20. Phillip Grouse, "It Does Not Compute," in Barry Thiering and Edgar Castle, eds., *Some Trust in Chariots* (New York: Popular Library, 1972), p. 118.
21. Landsburg, *In Search of Ancient Mysteries,* pp. 100–1.
22. Martin Gardner, *In the Name of Science* (New York: Putnam, 1952), pp. 174–181.

Chapter 2

1. Alan and Sally Landsburg, *In Search of Ancient Mysteries* (New York: Bantam Books, 1974), p. 66.
2. Gerald Hawkins, *Stonehenge Decoded* (New York: Doubleday, 1965).
3. Clifford Wilson, *Crash Go The Chariots* (New York: Lancer Books, 1972).
4. Andrew Tomas, *The Home of the Gods* (New York: Berkley, 1974), p. 65.
5. Peter Kolosimo, *Not of this World,* (New York: Bantam Books, 1973), p. 9.
6. Morris Goran, *The Future of Science* (New York: Spartan, 1971), Ch. 13.
7. Jack Stoneley and A.T. Lawton, *Is Anyone Out There?* (New York: Warner, 1974), p. 221.
8. Jacques Bergier, *Extra-Terrestrial Visitations From Pre-Historic Times to the Present* (Chicago: Regnery, 1973), p. 33–4.
9. *National Enquirer,* 24 December 1974, p. 4.
10. *Ibid.,* p. 4.
11. I.S. Shklovskii and Carl Sagan, *Intelligent Life in the Universe* (New York: Dell, 1974), pp. 454–5.
12. Donald Menzel, "UFO's—The Modern Myth" in C. Sagan and T. Page, eds., *UFOs—A Scientific Debate* (Ithaca: Cornell University Press, 1972), pp. 175–6.
13. Erich von Däniken, *Chariots of the Gods?* (New York: Bantam Books, 1971), p. 37.
14. Erich von Däniken, *The Gold of the Gods* (New York: Putnam, 1973), p. 7.

15. *Ibid.*, p. 106.
16. *Chicago Tribune*, 3 May 1974.
17. Bergier, *Extra-Terrestrial Visitations*, p. 45.
18. *Ibid.*, p. 47.
19. Kolosimo, *Not of This World*, p. 5.
20. W. Raymond Drake, *Gods and Spacemen in the Ancient West* (New York: New American Library, 1974), p. 43.
21. Erich von Däniken, "Foreword" in Max H. Flindt and Otto O. Binder, *Mankind—Child of the Stars* (New York: Fawcett, 1974), p. 11.
22. *Chicago Tribune*, p. 4, 3 May 1974.
23. Basil Hennessy, "Archaeology of the Ancient Near East" in Barry Thiering and Edgar Castle, eds., *Some Trust in Chariots*, (New York: Popular Library, 1972), p. 8.
24. *Playboy*, August 1974, p. 51.
25. *Ibid.*, p. 151.
26. *Ibid.*, p. 58.
27. *Ibid.*, p. 60.
28. Noel Weeks, "A Funny Thing Happened on the Way to Mesopotamia," in Barry Thiering and Edgar Castle, eds., *Some Trust in Chariots* (New York: Popular Library, 1972), p. 76.

Chapter 3

1. Immanuel Velikovsky, *Worlds in Collision* (New York: Macmillan, 1950).
2. George Grinnell, "Trying to Find the Truth About the Controversial Theories of Velikovsky," *Science Forum*, 7 April 1974, p. 4.
3. Lynn White, Jr., *Science, Scientists and Politics* (Santa Barbara, California: Center for the Study of Democratic Institutions, 1963), pp. 14–15.
4. Jean Sendy, *Those Gods Who Made Heaven and Earth* (New York: Berkley, 1970), p. 101.
5. Erich von Däniken, *Chariots of the Gods?* (New York: Bantam Books, 1971), p. 34.
6. *Ibid.*, p. 36.
7. Alan and Sally Landsburg, *In Search of Ancient Mysteries* (New York: Bantam Books, 1974), p. 142.
8. Josef F. Blumrich, *The Spaceships of Ezekiel* (New York: Bantam Books, 1974).
9. Donald Menzel, "UFO's—The Modern Myth" in C. Sagan and T. Page, eds., *UFOs—A Scientific Debate* (Ithaca, N.Y.: Cornell University Press, 1972), p. 213.

NOTES 161

10. Eric Norman, *Gods Demons and Space Chariots* (New York: Lancer Books, 1970), p. 24–31.
11. von Däniken, *Chariots of the Gods?*, p. 56.
12. *Playboy*, 21 August 1974, p. 51.
13. John Gent, "Riders to the Chariot" in Barry Thiering and Edgar Castle, eds., *Some Trust in Chariots* (New York: Popular Library, 1972), p. 92.
14. Robert Charroux, *Legacy of the Gods* (New York: Berkley, 1974), p. 99.
15. Erich von Däniken, *The Gold of the Gods* (New York: Putnam, 1973), pp. 47–8.
16. *Ibid.*, p. 47–8.
17. Samuel Rosenberg, "UFOs in History" in Daniel S. Gillmor, ed., *Scientific Study of Unidentified Flying Objects* (New York: Bantam Books, 1969), p. 243.
18. W. Raymond Drake, *Gods and Spacemen in the Ancient West* (New York: New American Library, 1974), p. 38.
19. W. Raymond Drake, *Gods and Spacemen in the Ancient East* (New York: Signet, 1973).
20. Drake, *Gods and Spacemen in West*, p. 126.
21. von Däniken, *Gold of the Gods*, pp. 180–83.
22. W. Raymond Drake, *Gods and Spacemen of the Ancient Past* (New York: New American Library, 1974), p. 13.
23. Sendy, *These Gods*, p. 69.
24. B.L. Trench, *The Sky People* (New York: Award Books, 1970), p. 86.
25. Brinsley Le Poer Trench, *Mysterious Visitors* (New York: Stein and Day, 1973), p. 138.
26. Drake, *Gods and Spacemen in West*, p. 178.
27. *Ibid.*, p. 46.
28. *Playboy*, 21 August 1974, p. 57.
29. *Chicago Tribune, The Trib*, Friday, 6 December 1974, p. 3.

Chapter 4

1. Jacques Bergier, *Extra-Terrestrial Visitations From Pre-Historic Times to the Present* (Chicago: Regnery, 1973).
2. Charles Fort, *Lo!* (New York: Ace Books, 1972), p. 156 ff.
3. Philip Morrison, "The Nature of Scientific Evidence: A Summary" in C. Sagan and T. Page, eds., *UFOs—A Scientific Debate* (Ithaca, N.Y.: Cornell University Press, 1972), p. 284.
4. Elliot Rose, *A Razor for A Goat* (Toronto: University of Toronto Press, 1962), p. 142.

5. Eric Norman, *Gods, Demons and Space Chariots* (New York: Lancer Books, 1970), pp. 86–89.
6. Andrew Tomas, *The Home of the Gods* (New York: Berkley, 1974), p. 113.
7. Peter Kolosimo, *Not of This World*, (New York: Bantam Books, 1973), p. 41.
8. Ivan T. Sanderson, *Uninvited Visitors* (New York: Cowles, 1967), p. 181.
9. *Ibid.*, pp. 198–9.
10. Norman, *Gods, Demons and Space Chariots*, pp. 16–17.
11. Erich von Däniken, *Gods From Outer Space* (New York: Bantam Books, 1971), p. 96.
12. Robert Charroux, *Legacy of the Gods* (New York: Berkley, 1974), pp. 293–301.
13. *Playboy*, 21 August 1974, p. 151.
14. Tomas, *The Home of the Gods*, p. 135.
15. *Ibid.*, p. 45.
16. W. Raymond Drake, *Gods and Spacemen in the Ancient West* (New York: New American Library, 1974), p. 46.
17. Martin Gardner, *In the Name of Science* (New York: Putnam, 1952), pp. 37–38.
18. Drake, *Gods and Spacemen in the West*, p. 5.
19. Kolosimo, *Not of This World*, p. 139.
20. Brinsley Le Poer Trench, *Mysterious Visitors* (New York: Stein and Day, 1973), pp. 18–19.
21. M. Goran, *Science and Anti-Science* (Ann Arbor, Michigan: Ann Arbor Science Publishers, 1974), ch. 7.
22. Michael A. Arbib, "The Likelihood of the Evolution of Communicating Intelligences on Other Planets" in Cyril Ponnamperuma and A.G. W. Cameron, eds., *Interstellar Communication: Scientific Perspectives* (Boston: Houghton Mifflin, 1974), p. 60.
23. Gardner, *In the Name of Science*, pp. 168–9.
24. John Godwin, *Occult America*, New York: Doubleday, 1972, p. 167.

Chapter 5

1. R.L. Dione, *God Drives A Flying Saucer* (New York: Bantam Books, 1973), p. vii.
2. Jean Sendy, *Those Gods Who Made Heaven and Earth* (New York: Berkley, 1970), p. 12.
3. Erich von Däniken, *Gods From Outer Space* (New York: Bantam Books, 1972), pp. 10–11.

4. Erich von Däniken, *Chariots of the Gods?* (New York: Bantam Books, 1971), p. 141.

5. von Däniken, *Gods From Outer Space*, pp. 2–3.

6. Jean Sendy, *The Coming of the Gods* (New York: Berkley, 1973), p. dedication.

7. Jacques Bergier, *Extra-Terrestrial Visitations from Pre-Historic Times To The Present*, (New York: New American Library, Signet, 1974), p. viii.

8. Andrew Tomas, *The Home of the Gods* (New York: Berkley, 1974), p. 138.

9. Philip H. Abelson, *Science*, 184, 21 June 1974.

10. Robert Charroux, *Legacy of the Gods* (New York: Berkley, 1974), p. 20.

11. *Ibid.*, p. 69.

12. Quoted in Max H. Flindt and Otto O. Binder, *Mankind, Child of The Stars* (New York: Fawcett, 1974), p. 16.

13. Erich von Däniken, *The Gold of the Gods* (New York: Putnam, 1973), pp. 68–9.

14. Flindt and Binder, *Mankind*, pp. 13–14.

15. Bergier, *Extra-Terrestrial Visitations*, pp. 117–138.

16. Brinsley Le Poer Trench, *Mysterious Visitors* (New York: Stein & Day, 1973), p. 112.

17. Bergier, *Extra-Terrestrial Visitations*, p. 126.

18. *Ibid.*, p. 103.

19. *Ibid.*, p. 68.

20. *Ibid.*, p. 86.

21. *Ibid.*, p. 8.

22. *Ibid.*, pp. 10–11.

23. Alan and Sally Landsburg, *In Search of Ancient Mysteries* (New York: Bantam Books, 1974), p. 161.

24. W. Raymond Drake, *Gods and Spacemen in the Ancient West* (New York: New American Library, 1974), p. 170.

25. Landsburg, *In Search of Ancient Mysteries*, p. 6.

26. *Ibid.*, p. 16.

27. *Ibid.*, pp. 16–17.

28. *Ibid.*, pp. 30–31.

29. *Ibid.*, p. 46.

30. *Ibid.*, pp. 87–8.

31. von Däniken, *The Gold of the Gods*, pp. 23–4.

32. Bergier, *Extra-Terrestrial Visitations*, p. 25.

33. Louis Pauwels and Jacques Bergier, *The Morning of the Magicians* (New York: Avon, 1968), p. 184.

34. Barbara J. Culliton, "The Sloan-Kettering Affair: A Story Without A Hero," *Science*, 184, 10 May 1974, pp. 644–650; and Barbara J. Culliton,

"The Sloan-Kettering Affair (II): An Uneasy Solution," *Science*, 184, 14 June 1974, pp. 1154–1157.
35. Robert Charroux, *The Gods Unknown* (New York: Berkley, 1974), p. xi.
36. *Playboy*, 21 August 1974, p. 57.

Chapter 6

1. W. Raymond Drake, *Gods and Spacemen in the Ancient West* (New York: New American Library, 1974), p. 3.
2. Eric Norman, *Gods, Demons and Space Chariots* (New York: Lancer Books, 1970), p. 115.
3. I.S. Shklovskii and Carl Sagan, *Intelligent Life in the Universe* (New York: Dell, 1974), p. 454.
4. Norman, *Gods, Demons, and Space Chariots*, p. 11.
5. *Ibid.*, p. 10.
6. Andrew Tomas, *The Home of the Gods* (New York: Berkley, 1974), p. 44.
7. Ralph Blum and Judy Blum, *Beyond Earth: Man's Contact With UFOs* (New York: Bantam Books, 1974), p. 25.
8. Max H. Flindt and Otto O. Binder, *Mankind, Child of the Stars* (New York: Fawcett, 1974), p. 52.
9. *Nature*, 251, 13 September 1974, p. 95.
10. William J. Kaufmann III, *Relativity and Cosmology* (New York: Harper & Row, 1972), p. 65.
11. Carl Sagan, ed., *Communication with Extraterrestrial Intelligence* (Cambridge: MIT Press, 1973), p. 186.
12. Carl Sagan, *The Cosmic Connection, An Extraterrestrial Perspective* (New York: Doubleday, 1973), p. 195.
13. John W. Macvey, *Whispers From Space* (New York: Macmillan, 1973), p. 237.
14. Ian Ridpath, "Long-Delayed Signals May Echo From Moon's Ghost," *New Scientist*, 64, 3 October 1974.
15. Tomas, *The Home of the Gods*, p. 93.
16. Jacques Bergier, *Extraterrestrial Visitations from Pre-Historic Times to the Present* (Chicago: Regnery, 1973), ch. 4.
17. Clifford Wilson, *Crash Go The Chariots* (New York: Lancer Books, 1972), p. 76.
18. A.D. Crown, "The South Sea Bubble Burst" in Barry Thiering and Edgar Castle, eds., *Some Trust in Chariots* (New York: Popular Library, 1972), pp. 27–8.

19. Leo Bagrow, *History of Cartography* (Cambridge: Harvard University Press, 1964), p. 87.
20. *Nature*, 184, 1959, p. 844.
21. Philip Morrison, "Conclusion: Entropy, Life, and Communication" in Cyril Ponnamperuna and A.G.W. Cameron, eds., *Interstellar Communication: Scientific Perspectives* (Boston: Houghton Mifflin, 1974), p. 169.
22. *Ibid.*, p. 171.
23. *Nature*, 252, 29 November 1974, p. 349.

Chapter 7

1. Howard S. Miller, "The Political Economy of Science" in George H. Daniels, ed., *Nineteenth-Century American Science, A Reappraisal* (Evanston: Northwestern University Press, 1972), p. 103.
2. Frank D. Drake, *Intelligent Life in Space* (New York: Macmillan, 1962), p. 37.
3. Project Cyclops, CR 114445, available from Dr. John Billingham, NASA/Ames Research Center, Code LT, Moffett Field, California, 94035, p. 1.
4. *Ibid.*, pp. 170–1.
5. Cyril Ponnamperuna and A.G.W. Cameron, eds., *Interstellar Communication: Scientific Perspectives* (Boston Houghton Mifflin, 1974).
6. Richard Berendzen, ed., *Life Beyond the Earth and the Mind of Man* (Washington: National Aeronautics and Space Administration, 1973).
7. Carl Sagan, ed., *Communication with Extraterrestrial Intelligence* (Cambridge, MIT Press, 1973).

Chapter 8

1. W. Raymond Drake, *Gods and Spacemen of the Ancient Past* (New York: New American Library, 1974), pp. 24–25.
2. J. Allen Hynek, *The UFO Experience* (New York: Ballantine, 1974), p. 161.
3. Eric Norman, *Gods, Demons and Space Chariots* (New York: Lancer Books, 1970), pp. 129–132.
4. R.L. Dione, *God Drives A Flying Saucer* (New York: Bantam Books, 1973), pp. 4–5.
5. Jacques Bergier, *Extra-Terrestrial Visitations From Pre-Historic Times to the Present* (Chicago: Regnery, 1973), pp. 146–7.

166 THE MODERN MYTH

6. W. Raymond Drake, *Gods and Spacemen in the Ancient West* (New York: New American Library, 1974), p. 99.

7. Brinsley Le Poer Trench, *Mysterious Visitors* (New York: Stein & Day, 1973), pp. 73–4, quoting Charles Bowen, ed., *The Humanoids* (Chicago: Regnery, 1970).

8. Drake, *Gods and Spacemen in the West*, p. 14.

9. Norman, *Gods, Demons and Space Chariots*, p. 6.

10. Dione, *God Drives*, p. 34.

11. Peter Kolosimo, *Not of This World* (New York: Bantam Books, 1973), p. 70.

12. Aimé Michel, *The Truth About Flying Saucers* (New York: Pyramid Books, 1974), p. 98.

13. *Ibid.*, p. 99.

14. Drake, *Gods and Spacemen in the West*, p. 14.

15. Ivan T. Sanderson, *Uninvited Visitors* (New York: Cowles Book, 1967), pp. 140–1.

16. Major Donald E. Keyhoe, *Aliens From Space* (New York: Doubleday, 1973), p. 19.

17. W. Raymond Drake, *Gods and Spacemen of the Ancient Past* (New York: New American Library, 1974), p. 7.

18. Max H. Flindt and Otto O. Binder, *Mankind, Child of the Stars* (New York: Fawcett, 1974), p. 238.

19. Keyhoe, *Aliens From Space*, pp. 23–6.

20. *Ibid.*, pp. 31–3.

21. *Ibid.*, p. 76.

22. John Godwin, *Occult America* (New York: Doubleday, 1972), pp. 145–6.

23. *Ibid.*, pp. 147–8.

24. *Ibid.*, p. 88.

25. Ralph and Judy Blum, *Beyond Earth: Man's Contact With UFO* (New York: Bantam Books), chapter 13, p. 143ff.

26. Trench, *Mysterious Visitors*, pp. 50–5.

27. Keyhoe, *Aliens From Space*, p. 227.

28. Hynek, *The UFO Experience* (New York: Ballantine, 1974), p. 164.

29. Donald Menzel, "UFOs: The Modern Myth" in C. Sagan and T. Page, eds., *UFOs—A Scientific Debate* (Ithaca: Cornell University Press, 1972), pp. 146–153.

30. Norman, *Gods, Demons and Space Chariots*, p. 177.

31. Philip J. Klass, *UFOs—Identified* (New York: Random House, 1968), chps. 20 and 21.

32. John G. Fuller, *The Interrupted Journey* (New York: Berkley, 1974), p. 319.

33. Ralph Blum, "Are UFOs For Real," *Reader's Digest*, June 1974, pp. 89–93; also Blum, *Beyond Earth*.
34. Klass, *UFOs—Identified*, chps. 18 and 19.
35. Trench, *Mysterious Visitors*, pp. 60–1.
36. Blum, *Beyond Earth*, pp. 179–80.
37. Martin Ebon, ed., *The Psychic Scene* (New York: Signet, 1974), p. 62.
38. Drake, *Gods and Spacemen of the West*, pp. 9–11.
39. Glenn McWane and David Graham, *The New UFO Sightings* (New York: Warner, 1974), pp. 116–32.
40. Trench, *Mysterious Visitors*, p. 115ff.
41. Christopher Evans, *Cults of Unreason* (London: Harrap, 1973), pp. 147–9.
42. Andrija Puharich, *Uri* (New York: Doubleday, 1974).
43. McWane and Graham, *UFO Sightings*, p. 113.
44. Robert Emenegger, *UFOs, Past, Present and Future* (New York: Ballantine, 1974), pp. 55–62.

Chapter 9

1. Harold T. Wilkins, *Flying Saucers on the Attack* (New York: Ace Books, 1967), p. 158.
2. George Adamski and Desmond Leslie, *Flying Saucers Have Landed* (New York: British Book Centre, 1953), p. 176.
3. Eric Norman, *Gods, Demons and Space Chariots* (New York: Lancer Books, 1970), pp. 132–3.
4. Christopher Evans, *Cults of Unreason* (London: Harrap, 1973), p. 143.
5. F.W. Holiday, *Creatures From The Inner Sphere* (New York: Popular Library, 1973), p. 181.
6. *Ibid.*, pp. 12–13.
7. Jacques Vallee, *Passport to Magonia* (Chicago: Regnery, 1969), p. 5.
8. Gordon I.R. Lore, Jr. and Harold H. Deneault, Jr., *Mysteries of the Skies: UFOs in Perspective* (Englewood Cliffs, N.J.: Prentice-Hall, 1968), p. 41.
9. *Science*, 160, 14 June 1968, p. 1260.
10. Aimé Michel, *The Truth About Flying Saucers* (New York: Pyramid Books, 1974), p. 31.
11. Andrew Tomas, *The Home of the Gods* (New York: Berkley, 1974), p. 46.
12. J. Allen Hynek, *The UFO Experience* (New York: Ballantine, 1974), p. 41.
13. Frank B. Salisbury, "The Scientist and the UFOs" in *Science and*

Mechanics Editors, ed., *The Official Guide to UFOs* (New York: Ace Books, 1968), pp. 44–66.

14. John G. Fuller, *Incident at Exeter* (New York: Berkley, 1974).

15. Hynek, *The UFO Experience*, p. 17.

16. Robert Emenegger, *UFOs, Past, Present and Future* (New York: Ballantine, 1974), pp. 12–15 and Max H. Flindt and Otto O. Binder, *Mankind, Child of the Stars* (New York: Fawcett, 1974), p. 239.

17. Emenegger, *UFOs, Past, Present and Future*, pp. 17–18.

18. Peter Kolosimo, *Not of This World* (New York: Bantam Books, 1973), p. 76.

19. Shiela Ostrander and Lynn Schroeder, *Psychic Discoveries Behind the Iron Curtain* (New York: Bantam Books, 1971), pp. 94–103.

20. Mort Young, *UFO: Top Secret* (New York: Essandess Special Editions, 1967), pp. 106–111.

21. *Ibid.*, pp. 30–31.

22. James E. McDonald, "Science In Default: Twenty-Two Years of Inadequate UFO Investigations" in C. Sagan and T. Page, eds., *UFOs—A Scientific Debate* (Ithaca: Cornell University Press, 1972), pp. 52–122.

23. *New Scientist*, 61, 28 March 1974, p. 832.

24. Glenn McWane and David Graham, *The New UFO Sightings* (New York: Warner, 1974), p. 65.

25. Emenegger, *UFOs, Past, Present and Future*, pp. 11–12.

26. McWane and Graham, *The New UFO Sightings*, p. 8.

Chapter 10

1. Renato Vesco, *Intercept UFO* (New York: Zebra Books, 1974), p. 37.

2. *Chicago Tribune*, 18 October 1973.

3. Vesco, *Intercept UFO*, p. 7.

4. Frank Scully, *Behind the Flying Saucers* (New York: Holt, 1950).

5. Curtis D. MacDougall, *Hoaxes* (New York: Dover, 1958), pp. 229–30.

6. Frank D. Drake, "On the Abilities and Limitations of UFOs and Similar Phenomena" in C. Sagan and T. Page, eds., *UFOs—A Scientific Debate* (Ithaca: Cornell University Press, 1972), p. 254.

7. Christopher Evans, *Cults of Unreason* (London: Harrap, 1973), pp. 144–5.

8. Glenn McWane and David Graham, *The New UFO Sightings* (New York: Warner, 1974), p. 98.

9. Ray Bradbury et al, *Mars and the Mind of Man* (New York: Harper & Row, 1973), pp. 13–14.

10. *Ibid.*, p. 6.

11. Richard Baum, *The Planets, Some Myths and Realities* (New York: John Wiley, 1973).

12. Philip Morrison, "The Nature of Scientific Evidence: A Summary", in C. Sagan and T. Page, eds., *UFOs—A Scientific Debate* (Ithaca: Cornell University Press, 1972), pp. 285–6.

13. Daniel McCaughna, "UFO Craze Foiled in Ohio," *Chicago Tribune*, 18 October 1973.

14. *Ibid.*

15. *Chicago Daily News*, 19 November 1973.

16. Philip Klass, *UFOs Explained* (New York: Random House, 1974), pp. 271, 18, 41, 90.

Chapter 11

1. Philip Morrison, "The Nature of Scientific Evidence: A Summary" in C. Sagan and T. Page, eds., *UFOs—A Scientific Debate* (Ithaca: Cornell University Press, 1972), p. 287.

2. William R. Corliss, *Strange Phenomena* (Glen Arm, Maryland: Box 107 C, 21057, Case GLD–033, 1974).

3. Donald Menzel, "UFOs: The Modern Myth" in Sagan and Page, eds., *UFOs—A Debate*, pp. 142–3.

4. Donald Menzel, *Flying Saucers* (Cambridge: Harvard University Press, 1953), p. vi.

5. *Argosy*, 379, April 1974, p. 86.

6. *Physics Today*, March 1974, p. 15.

7. Glenn McWane and David Graham, *The New UFO Sightings* (New York: Warner, 1974), p. 28.

8. Vincent Gaddis, *Invisible Horizons* (New York: Ace Books, 1965), pp. 224–236.

9. Philip J. Klass, *UFOs—Identified* (New York: Random House, 1968) and *UFOs Explained* (New York: Random House, 1974).

10. J. Dale Berry, *Journal of Atmospheric and Terrestrial Physics*, 29, pp. 1095–1101, Nov. 1967.

11. R. Vesco, *Intercept UFO* (New York: Zebra Books, 1974), p. 112.

12. Ivan T. Sanderson, *Uninvited Visitors* (New York: Cowles, 1967), p. 101.

13. Martin Gardner, *In The Name of Science* (New York: Putnam, 1952), p. 65.

14. Sanderson, *Uninvited Visitors*, p. 182.

15. *Ibid.*, p. 208.

16. R.L. Dione, *God Drives A Flying Saucer* (New York: Bantam Books, 1973), p. 12.

17. Brinsley Le Poer Trench, *Mysterious Visitors* (New York: Stein & Day, 1973), p. 11.

18. F.W. Holiday, *Creatures From the Inner Sphere* (New York: Popular Library, 1973), p. 214.

19. Eric Norman, *Gods, Demons and Space Chariots* (New York: Lancer Books, 1970), pp. 201–5.

20. Brinsley Le Poer Trench, *The Sky People* (New York: Award Books, 1970), p. 151.

21. Trench, *Mysterious Visitors*, p. 71.

22. Donald Keyhoe, *The Flying Saucers Are Real* (New York, Fawcett, 1950); *Flying Saucers From Outer Space* (New York: Holt, 1953); *Flying Saucer Conspiracy* (New York: Holt, 1955).

23. D.H. Rawcliffe, *Illusions and Delusions of the Supernatural and Occult* (New York: Dover, 1959), p. 178ff.

24. Nandor Fodor, *Freud, Jung and Occultism* (New Hyde Park, N.Y.: University Books, 1971), p. 14.

25. M.K. Jessup, *The Case for the UFO* (New York: Citadel Press, 1955).

26. Graham Chedd, "Colonisation at Lagrangia," *New Scientist*, 64, 24 October 1974, pp. 247–9.

27. Sanderson, *Uninvited Visitors*, p. 137.

28. W. Raymond Drake, *Gods and Spacemen in the Ancient West* (New York: New American Library, 1974), p. 5.

29. *Ibid.*, pp. 8–9.

30. Carl Sagan, "UFOs: The Extraterrestrial and Other Hypotheses," Sagan and Page, eds., *UFOs—A Debate*, p. 271.

31. Andrija Puharich, *Beyond Telepathy* (London: Darton Longman and Todd, 1962), pp. 41–2.

32. Gordon H. Evans, "UFO: Theories of Flight" in Editors of *Science and Mechanics*, eds., *The Official Guide To UFOs* (New York: Ace Books, 1968), pp. 9–10.

33. Aimé Michel, *The Truth About Flying Saucers* (New York: Pyramid Books, 1974), pp. 210–26.

34. *Ibid.*, p. 214.

35. Jacques Bergier, *Extraterrestrial Visitations from Pre-Historic Times to the Present* (Chicago: Regnery, 1973), p. 184.

36. Menzel, "UFOs: The Modern Myth," pp. 75–6, 189–205.

37. Major Donald E. Keyhoe, *Aliens From Space* (New York: Doubleday, 1973), p. 211.

38. *Newsweek*, 20 January 1975, p. 72.

Chapter 12

1. Major Donald E. Keyhoe, *Aliens From Space* (New York: Doubleday, 1973), pp. 306–7.
2. Robert Loftin, *Identified Flying Saucers* (New York: David McKay, 1968), pp. 2–3.
3. Max Flindt and Otto Binder, *Mankind, Child of the Stars* (New York: Fawcett, 1974), p. 233.
4. Mort Young, *UFO: Top Secret* (New York: Essandess Special Editions, 1967), pp. 41–62.
5. Keyhoe, *Aliens From Space*, p. 92.
6. Loftin, *Identified Flying Saucers*, pp. 107–8.
7. David R. Saunders and R. Rogers Harkins, *"UFOs Yes!—Where The Condon Committee Went Wrong* (New York: Signet, 1968).
8. Jacques Vallee, *Anatomy of A Phenomenon* (Chicago: Regnery, 1965) and Jacques and Janine Vallee, *Challenge to Science: The UFO Enigma* (Chicago: Regnery, 1966).
9. Carl Sagan and T. Page, eds., *UFOs—A Scientific Debate* (Ithaca: Cornell University Press, 1972; also W.W. Norton, 1974).
10. Harvey H. Nininger, *Find A Falling Star* (New York: Paul S. Eriksson, 1972), pp. 4–5.
11. J. Allen Hynek, *The UFO Experience* (New York: Ballantine, 1974), p. 193.
12. *Ibid.*, p. 8.
13. *Ibid.*, p. 233.
14. William Markowitz, "The Physics and Metaphysics of Unidentified Flying Objects," *Science*, 157, 15 September 1967, pp. 1274–1279.
15. Carl Sagan, *The Cosmic Connection* (New York: Doubleday, 1973), p. 203.
16. *Science News*, 3 November 1973, p. 284.
17. Loftin, *Identified Flying Saucers*, p. 1.
18. Young, *UFO: Top Secret*, p. 132.
19. Keyhoe, *Aliens From Space*, p. 46.
20. Kurt Seligmann, *The Mirror of Magic* (New York: Pantheon, 1948), pp. 458–9.
21. *Chicago Tribune Magazine*, 9 June 1974, pp. 12–14.
22. Ivan T. Sanderson, *Uninvited Visitors* (New York: Cowles, 1967), p. 163.
23. Loftin, *Identified Flying Saucers*, pp. 16–20.
24. Frank Scully, *Behind the Flying Saucers* (New York: Holt, 1950).
25. W. Raymond Drake, *Gods and Spacemen in the Ancient West* (New York: New American Library, 1974), p. 6.

Bibliography

Selected Readings—Ancient Astronauts

Aylesworth, Thomas. *Who's Out There?* New York: McGraw-Hill, 1975.

Baxter, John, and Atkins, Thomas. *The Fire Came By.* Garden City, N.Y.: Doubleday, 1976.

Berendzen, Richard, ed. *Life Beyond the Earth and the Mind of Man.* Washington, D.C.: National Aeronautics and Space Administration, 1973.

Bergier, Jacques, and INFO editors. *Extraterrestrial Intervention: The Evidence.* Chicago: Regnery, 1974.

Bergier, Jacques. *Extraterrestrial Visitations From Prehistoric Times to the Present.* Chicago: Regnery, 1973; New York: New American Library (Signet), 1974.

Berlitz, Charles. *Mysteries From Forgotten Worlds.* New York: Doubleday, 1972; New York: Dell Laurel, 1973.

Blumrich, Josef F. *The Spaceships of Ezekiel.* New York: Bantam Books, 1974.

Charroux, Robert. *Legacy of the Gods.* New York: Berkley Medallion, 1974.

Charroux, Robert. *Masters of the World.* New York: Berkley Medallion, 1974.

Charroux, Robert. *One Hundred Thousand Years of Man's Unknown History.* New York: Berkley Medallion, 1974.

Charroux, Robert. *The Gods Unknown.* New York: Berkley Medallion, 1974.

Churchward, James. *The Children of Mu.* London: Neville Spearman, 1959.

Churchward, James. *The Lost Continent of Mu.* London: Neville Spearman, 1959.

Cohen, Daniel. *The Ancient Visitors.* New York: Doubleday, 1976.

Collyns, Robin. *Did Spacemen Colonize the Earth?* Chicago: Regnery, 1976.

Der Spiegel, editors. "Anatomy of A World Best-Seller," *Encounter,* Aug. 1973, XLI, pp. 8–17.

Dione, R.L. *God Drives a Flying Saucer.* New York: Bantam Books, 1973.

Donnelly, Ignatius. *Atlantis, The Antediluvian World.* London: Sigwick and Jackson, 1950.

Drake, W. Raymond. *Gods and Spacemen in the Ancient East.* New York: New American Library (Signet), 1973.

Drake, W. Raymond. *Gods and Spacemen in History.* Chicago: Regnery, 1975.

Drake, W. Raymond. *Gods and Spacemen of the Ancient Past.* New York: New American Library (Signet), 1974.

Drake, W. Raymond. *Gods and Spacemen in the Ancient West.* New York: New American Library (Signet), 1974.

Drake, Frank D. *Intelligent Life in Space.* New York: Macmillan, 1962.

Flindt, Max H. and Binder, Otto O. *Mankind—Child of the Stars.* Greenwich, Connecticut: Fawcett Gold Medal, 1974.

Fort Charles. *Lo!.* New York: Holt, Rinehart and Winston, 1941; New York: Ace Books, 1972.

Fort, Charles. *New Lands.* New York: Holt, Rinehart and Winston, 1941; New York: Ace Books, 1972.

Fort, Charles. *The Book of the Damned.* New York: Holt, Rinehart and Winston, 1941; New York: Ace Books, 1972.

Fort, Charles. *Wild Talents.* New York: Holt, Rinehart and

Winston, 1941; New York: Ace Books, 1972.

Ginsbergh, Irwin. *First Man, Then Adam.* Chicago: Dearborn Press, 1975.

Hapgood, C.H. *Map of the Ancient Sea Kings.* Radner, Pennsylvania: Chilton Books, 1965.

Hawkins, Gerald. *Stonehenge Decoded.* Garden City, New York: Doubleday, 1965.

Heyerdahl, Thor. *Aku-Aku, the Secret of Easter Island.* London: George Allen and Unwin, 1958.

Hitching, Francis. *Earth Magic.* New York: Morrow, 1977.

Holiday, F.W. *Creatures from the Inner Sphere.* New York: Popular Library, 1973.

Keel, John A. *Our Haunted Planet.* Greenwich, Connecticut: Fawcett, 1971.

Kolosimo, Peter. *Not of This World.* Secaucus, New Jersey: University Books, 1971; New York: Bantam Books, 1973.

Kolosimo, Peter. *Timeless Earth.* Secaucus, New Jersey: University Books, 1973.

Landsburg, Alan and Sally. *In Search of Ancient Mysteries.* New York: Bantam Books, 1974.

Landsburg, Alan and Sally. *The Outer Space Connection.* New York: Bantam Books, 1975.

Landsburg, Alan. *In Search of Extra-Terrestrials.* New York: Bantam, 1976.

Landsburg, Alan. *In Search of Lost Civilizations.* New York: Bantam, 1976.

Macvey, John W. *Whispers from Space.* New York: Macmillan, 1973.

Mooney, Richard E. *Colony: Earth.* New York: Stein and Day, 1974; Greenwich, Connecticut: Fawcett Crest, 1975.

Mooney, Richard E. *Gods of Air and Darkness.* New York: Stein and Day, 1976; Greenwich, Connecticut: Fawcett Crest, 1976.

Navia, Luis E. *A Bridge to the Stars.* Wayne, N.J.: Avery, 1977.

Norman, Eric. *Gods, Demons and Space Chariots.* New York: Lancer Books, 1970.

Ossendowski, F. *Beasts, Men and Gods.* New York: Dutton, 1926.

Pauwels, Louis and Bergier, Jacques. *The Morning of the Magicians.* New York: Stein and Day, 1964; New York: Avon Books, 1968.

Playboy Magazine. "Interview with Erich von Däniken." August 1974, p. 51.

Ponnamperuma, Cyril and Cameron, A.G.W., eds. *Interstellar Communication: Scientific Perspectives.* Boston: Houghton Mifflin, 1974.

Project Cyclops, CR 114445, from Dr. John Billingham, NASA/ Ames Research Center, Code LT. Moffett Field, California, 94035.

Sagan, Carl, ed. *Communication with Extraterrestrial Intelligence.* Cambridge, Massachusetts: MIT Press, 1973.

Sagan, Carl. *The Cosmic Connection.* New York: Doubleday, 1973, Dell, 1975.

Sanderson, Ivan T. *Uninvited Visitors.* New York: Cowles, 1967.

Sendy, Jean. *The Coming of the Gods.* New York: Berkley Medallion, 1973.

Sendy, Jean. *Those Gods Who Made Heaven and Earth.* New York: Berkley Medallion, 1970.

Shklovskii, I.S. and Sagan, Carl. *Intelligent Life in the Universe.* New York: Dell, 1974.

Sitchin, Z. *The Twelfth Planet.* New York: Stein and Day, 1976.

Steiger, Brad and White, John, eds. *Other Worlds, Other Universes.* New York: Doubleday, 1975.

Story, Ronald. *The Space-Gods Revealed.* New York: Harper and Row, 1976.

Temple, Robert K.G. *The Sirius Mystery.* London: Sidgwick and Jackson, 1976; New York: St. Martins, 1976.

Thiering, Barry and Castle, Edgar, eds. *Some Trust in Chariots.* New York: Popular Library, 1972.

Thom, Alexander. *Megalithic Lunar Observations.* Clarendon: Oxford University Press, 1971.

Thomas, Paul. *Flying Saucers Through the Ages.* London: Neville Spearman, 1970.

Tomas, Andrew. *The Home of the Gods.* New York: Berkley Medallion, 1974.

Tomas, Andrew. *We Are Not the First.* New York: Putnam, 1971; New York: Bantam Books, 1973.

Trench, Brinsley Le Poer. *Temple of the Stars.* New York: Ballantine Books, 1974.

Trench, Brinsley Le Poer. *The Sky People.* New York: Award Books, 1970.

von Däniken, Erich. *Chariots of the Gods?.* New York: Putnam, 1970; New York: Bantam Books, 1971.

von Däniken, Erich. *Gods From Outer Space.* New York: Putnam, 1971; New York: Bantam Books, 1972.

von Däniken, Erich. *In Search of Ancient Gods.* New York: Putnam, 1974.

von Däniken, Erich. *Miracles of the Gods.* New York: Delacorte, 1975.

von Däniken, Erich. *The Gold of the Gods.* New York: Putnam, 1973; New York: Bantam Books, 1974.

Watson, Lyall. *Supernature.* New York: Doubleday, 1973.

Wernick, Robert. *The Monument Builders.* New York: Time-Life Books, 1973.

Wilkins, Harold T. *Mysteries of Ancient South America.* Secaucus, N.J.: Citadel, 1974.

Williamson, George Hunt. *Road in the Sky.* London: Neville Spearman, 1968.

Wilson, Clifford. *Crash Go the Chariots.* New York: Lancer Books, 1972.

Wilson, Clifford. *The Chariots Still Crash.* New York: New American Library (Signet), 1976.

Wingert, Paul. *Primitive Art.* Clarendon: Oxford University Press, 1962.

Selected Readings—UFOs

Adamski, George and Leslie, Desmond. *Flying Saucers Have Landed.* New York: British Book Centre, 1953.

Adamski, George. *Flying Saucers Farewell.* New York: Abelard-Schuman, 1961.

Adamski, George. *Behind the Flying Saucer Mystery.* New York: Paperback Library, 1967.

Adamski, George. *Inside the Flying Saucers.* New York: Paperback Library, 1967.

Adamski, George. *Inside the Spaceships.* New York: Abelard-Schuman, 1955.

Angelucci, Orfeo F. *The Secret of the Saucers.* Amherst, Wisconsin: Amherst Press, 1955.

Arnold, Kenneth and Palmer, Ray. *The Coming of the Saucers.* Amherst, Wisconsin: Amherst Press, 1952.

Barker, Gray. *They Knew Too Much About Flying Saucers.* New York: Tower, 1967.

Bender, Albert K. *Flying Saucers and the Three Men.* New York: Paperback Library, 1968.

Bieri, Robert. "Humanoids on Other Planets." *American Scientist,* 52, Dec. 1964, pp. 452–458.

Binder, Otto. *Flying Saucers Are Watching Us.* New York: Belmont, 1968.

Binder, Otto. *What We Really Know About Flying Saucers.* Greenwich, Connecticut: Fawcett, 1967.

Binder, Otto. *Unsolved Mysteries of the Past.* New York: Tower, 1966.

Blum, Ralph and Judy. *Beyond Earth: Man's Contact with UFOs.* New York: Bantam Books, 1974.

Bowen, Charles. *The Humanoids.* London: Neville Spearman, 1969.

Burt, Eugene. *UFOs and Diamagnetism.* New York: Exposition Press, 1970.

Catoe, Lynn E. *UFOs and Related Subjects, An Annotated*

Bibliography. Washington: U.S. Government Printing Office, 1969.

Downing, Barry H. *The Bible and Flying Saucers.* Philadelphia: Lippincott, 1968; New York: Avon, 1970.

Eden, Jerome. *Planet in Trouble—The UFO Assault on Earth.* Hicksville, New York: Exposition, 1973.

Edwards, Frank. *Flying Saucers, Here and Now!.* New York: Lyle Stuart, 1967.

Edwards, Frank. *Flying Saucers, Serious Business.* New York: Lyle Stuart, 1966.

Emenegger, Robert. *UFO's Past, Present and Future.* New York: Ballantine, 1974.

Festinger, Riecken and Schacter. *When Prophecy Fails.* Minneapolis: University of Minnesota, 1956.

Flammonde, Paris. *The Age of Flying Saucers.* New York: Hawthorn, 1971.

Flammonde, Paris. *UFOs Exist!.* New York: Putnam, 1976.

Fry, Daniel W. *The White Sands Incident.* Louisville: Best Books, 1966.

Fuller, John G. *Aliens in the Sky.* New York: Putnam, 1969.

Fuller, John G. *Incident at Exeter.* New York: Putnam, 1966; New York: Berkley Medallion, 1974.

Fuller, John G. *The Interrupted Journey.* New York: Dial, 1966; New York: Berkley Medallion, 1974.

Gaddis, Vincent. *Invisible Horizons.* New York: Ace Books, 1965.

Gilfillan, Edward S., Jr. *Migration to the Stars.* New York: Robert B. Luce, 1975.

Gillmor, Daniel S., ed. *Scientific Study of Unidentified Flying Objects.* New York: Dutton, 1969; New York: Bantam Books, 1969.

Hearings, Committee on Science and Astronautics, U.S. House of Representatives, 90th Congress, 2nd Session. Symposium on Unidentified Flying Objects, July 1968. Washington: U.S. Government Printing Office, 1968.

Heard, Gerald. *The Riddle of Flying Saucers*. London: Carroll and Nicholson, 1950.

Hersey, Michael. *UFOs—The American Scene*. New York: St. Martins, 1976.

Hobana, Jon and Weverbaugh, Julien. *UFOs From Behind the Iron Curtain*. New York: Bantam Books, 1975.

Holzer, Hans. *The UFO-Nauts*. Greenwich, Connecticut: Fawcett, 1976.

Hynek, J. Allen and Vallee, Jacques. *The Edge of Reality*. Chicago: Regnery, 1975.

Hynek, J. Allen. *The UFO Experience*. Chicago: Regnery, 1972; New York: Ballantine, 1974.

Jacobs, David M. *The UFO Controversy in America*. Bloomington, Indiana: Indiana University Press, 1975; New York: New American Library, 1976.

Jessup, Morris K. *The Case for the UFOs*. New York: Citadel, 1955.

Jessup, Morris K. *The Expanding Case for the UFOs*. New York: Citadel, 1957.

Jung, Carl G. *Flying Saucers*. New York: Harcourt Brace World, 1959; New York: New American Library (Signet), 1969.

Keel, John A. *Strange Creatures From Time and Space*. Greenwich, Conn.: Fawcett, 1970.

Keel, John A. *Jadoo*. New York: Pyramid, 1972.

Keel, John A. *The Mothman Prophecies*. New York: Saturday Review Press, 1975.

Keel, John A. *UFOs, Operation Trojan Horse*. London: Sphere Books, 1973.

Keyhoe, Donald E. *Flying Saucers Are Real*. Greenwich, Connecticut: Fawcett, 1950.

Keyhoe, Donald E. *Flying Saucer Conspiracy*. New York: Holt, 1955.

Keyhoe, Donald E. *Flying Saucers From Outer Space*. New York: Holt, 1953; New York: Universal-Tandem, 1970.

Keyhoe, Donald E. *Flying Saucer: Top Secret.* New York: Putnam, 1960.

Klass, Philip J. *UFOs Explained.* New York: Random House, 1974; 1976.

Klass, Philip J. *UFOs Identified.* New York: Random House, 1968.

Loftin, Robert. *Identified Flying Saucers.* New York: David McKay, 1968.

Lore, Gordon I.R., Jr. and Deneault, Harold H., Jr. *Mysteries of the Skies: UFOs in Perspective.* Englewood Cliffs, New Jersey: Prentice-Hall, 1968.

Lorenzen, Coral and James. *Encounters with UFO Occupants.* New York: Berkley, 1976.

Lorenzen, Coral and James. *Flying Saucer Occupants.* New York: New American Library (Signet), 1968.

Lorenzen, Coral. *Flying Saucers, The Startling Evidence of the Invasion from Outer Space.* New York: New American Library (Signet), 1966.

Lorenzen, Coral. *The Great Flying Saucer Hoax.* New York: William Frederick Press, 1962.

Lorenzen, Coral and James. *UFOs Over the Americas.* New York: New American Library (Signet), 1968.

Lorenzen, Coral and James. *UFOs—The Whole Story.* New York: New American Library (Signet), 1969.

McCampbell, James M. *Ufology.* Belmont, Calif.: Jaymac, 1973; Millbrae, Calif.: Celestial Arts, 1976.

McWane, Glenn and Graham, David. *The New UFO Sightings.* New York: Warner Paperback, 1974.

Menzel, Donald H. and Boyd, Lyle. *The World of Flying Saucers.* New York: Doubleday, 1963.

Menzel, Donald H., and Tave, Ernest H. *The UFO Enigma.* New York: Doubleday, 1977.

Michel Aimé. *Flying Saucers and the Straight-Line Mystery.* New York: Criterion Books, 1958.

Michel, Aimé. *The Truth About Flying Saucers.* New York: Pyramid, 1974.

Palmer, Raymond A. *The Real UFO Invasion.* San Diego: Greenleaf, 1967.

Ruppelt, Edward J. *The Report on Unidentified Flying Objects.* New York: Doubleday, 1956.

Sagan, Carl and Page, T., eds., *UFOs—A Scientific Debate.* Ithaca, New York: Cornell University Press, 1972; New York: W.W. Norton, 1974.

Salisbury, Frank B. *The Utah UFO Display.* Greenwich, Connecticut: Devin-Adair, 1974.

Saunders, Davis R. and Harkins, R. Roger. *UFOs? Yes!.* New York: New American Library (Signet), 1968.

Science and Mechanics editors. *The Official Guide to UFOs.* New York: Ace Books, 1968.

Scully, Frank. *Behind the Flying Saucers.* New York: Holt, 1950.

Spencer, John Wallace. *Limbo of the Lost.* New York: Bantam Books, 1973.

Spencer, John Wallace. *No Earthly Explanation.* New York: Bantam Books, 1975.

Steiger, Brad, ed. *Project Bluebook.* New York: Ballantine, 1976.

Steiger, Brad. *The Gods of Aquarius.* New York: Harcourt Brace Jovanovich, 1976.

Steiger, Brad with Whritenour, Joan. *Allende Letters—New UFO Breakthrough.* New York: Award Books, 1968.

Steiger, Brad. *Mysteries of Time and Space.* Englewood Cliffs, New Jersey: Prentice-Hall, 1974.

Steiger, Brad and White, John, eds. *Other Worlds, Other Universes.* New York: Doubleday, 1975.

Stoneley, Jack and Lawton, A.T. *Is Anyone Out There?.* New York: Warner Paperback, 1974.

Tacker, Lawrence J. *Flying Saucers and the U.S. Air Force.* Princeton, New Jersey: Van Nostrand, 1960.

Trench, Brinsley Le Poer. *The Flying Saucer Story*. New York: Ace Books, 1966.

Vallee, Jacques. *Anatomy of a Phenomenon*. Chicago: Regnery, 1965; New York: Ballantine, 1974.

Vallee, Jacques. *Challenge to Science, The UFO Enigma*. Chicago: Regnery, 1966; New York: Ballantine, 1974.

Vallee, Jacques. *Passport to Magonia*. Chicago: Regnery, 1969.

Vallee, Jacques. *The Invisible College*. New York: Dutton, 1976.

Vesco, Renato. *Intercept UFO*. New York: Zebra Books, 1974.

Weldon, John and Levitt, Zola. *UFOs—What on Earth is Happening*. Los Angeles: Harvest House, 1975; New York: Bantam, 1976.

White, Dale. *Is Something Up There?*. New York: Scholastic Books, 1968.

Wilkins, Harold. *Flying Saucers on the Attack*. New York: Ace Books, 1967.

Wilson, Clifford. *UFOs and Their Mission Impossible*. New York: New American Library (Signet), 1974.

Wilson, Don. *Our Mysterious Spaceship Moon*. New York: Dell, 1976.

Young, Mort. *UFO Top Secret*. New York: Essandess Special Edition, 1967.

Periodicals Regularly Reporting on UFOs

Argosy UFO
Beyond Reality
Cosmic Frontiers
Dell's UFO Reports
Fate
Flying Saucer and UFO Quarterly
Flying Saucer News
Flying Saucer Review
Flying Saucers Magazine

INFO
Inforespace
International UFO Reporter
Mufon
Male
National Enquirer
Official UFO; also publishes *Ancient Astronaut*
Probe
Saga; also publishes *UFO Report* and *Annual UFO*
Saucer News
Tattler
True
UFO Investigator
UFOlogy

Index

185